T0093757

THE CLASSICAL
DOUBLE COPY

New Connections in
Gauge Theory and Gravity

THE CLASSICAL DOUBLE COPY

New Connections in
Gauge Theory and Gravity

Chris D. White

Queen Mary University of London, UK

World Scientific

NEW JERSEY · LONDON · SINGAPORE · BEIJING · SHANGHAI · HONG KONG · TAIPEI · CHENNAI · TOKYO

Published by

World Scientific Publishing Europe Ltd.

57 Shelton Street, Covent Garden, London WC2H 9HE

Head office: 5 Toh Tuck Link, Singapore 596224

USA office: 27 Warren Street, Suite 401-402, Hackensack, NJ 07601

Library of Congress Cataloging-in-Publication Data
Names: White, Chris D., author.
Title: The classical double copy : new connections in gauge theory and gravity /
    Chris D. White, Queen Mary University of London, UK.
Description: New Jersey : World Scientific, [2024] | Includes bibliographical references and index.
Identifiers: LCCN 2024002473 | ISBN 9781800615458 (hardcover) |
    ISBN 9781800615465 (ebook) | ISBN 9781800615472 (ebook other)
Subjects: LCSH: Gauge fields (Physics) | Gravitation. | General relativity (Physics)
Classification: LCC QC793.3.G38 W45 2024 | DDC 530.14/35--dc23/eng/20240226
LC record available at https://lccn.loc.gov/2024002473

British Library Cataloguing-in-Publication Data
A catalogue record for this book is available from the British Library.

For any available supplementary material, please visit
https://www.worldscientific.com/worldscibooks/10.1142/Q0457#t=suppl

Desk Editors: Nambirajan Karuppiah/Ana Ovey/Shi Ying Koe

Typeset by Stallion Press
Email: enquiries@stallionpress.com

To my brother Martin and sister Nicola.

Preface

In recent years, a remarkable new correspondence has emerged between our various theories of fundamental physics. Of the four fundamental forces in nature, three are described by so-called (non-) abelian gauge theories, a type of quantum field theory whose intricate mathematical structure was first uncovered in the latter half of the twentieth century. The fourth force, gravity, is best described by Einstein's general theory of relativity, which has stubbornly resisted efforts to make it quantum. Candidate theories such as string theory give us crucial insights, but it remains unclear how our actual universe emerges in the low-energy limit.

As a consequence of the above dichotomy, traditional methods of calculation in gauge theories and gravity look completely different largely due to the fact that they are applied by different groups of people, with their own distinct research groups and conferences. A huge amount is known about gauge theories due to their immediate application to particle accelerator experiments, such as the Large Hadron Collider. Likewise, general relativity underlies astrophysics and cosmology, including the recent discovery of gravitational waves.

Just over a decade ago at the time of writing, it was found that certain quantities in quantum field theory – the *scattering amplitudes* that underlie particle collision processes – can be cast in a form that looks almost identical in gauge theory and gravity. This correspondence became known as the *double copy* and was itself inspired by previous work in string theory. However, the double copy in field theory goes much further and has also been extended to show how classical solutions in gravity can be obtained from (simpler) gauge theory results. Besides connecting theories, the double copy implies the existence of intriguing new symmetry structures in gauge theory, which force us to think about these theories in a new way. Since its original inception, the web of theories connected by double-copy-like relationships has continually increased, such that it genuinely feels as if the foundations of quantum field theory have to be rewritten in order to make manifest a common underlying structure that, up to now, has remained largely hidden. There is also a highly practical motivation for studying the double copy: it tells us that complicated classical calculations in general relativity needed for astrophysics can be massively simplified by recycling methods and results from gauge theory. This in turn has created an intense and fruitful conversation between researchers in high energy and astrophysics, which has already resulted in cutting-edge calculations that would not have been previously possible.

Given the increasingly interdisciplinary nature of this subject, I felt it timely to write this book, with the aim of making the double copy broadly accessible to a wide spectrum of scientists, including physicists, mathematicians and astronomers. It seems after all likely that the coming years may see a dialogue involving an even broader range of subfields, including optics, condensed matter physics and pure mathematics, as we will describe later on. A problem then arises in that the double copy connects the two large subjects of quantum field theory and general relativity. Both of these subjects are highly technical and not necessarily part of the standard core knowledge of the average physicist. The challenge is then to explain the double copy and its related ideas by relying on basic notions from these theories without the higher knowledge possessed by a contemporary researcher. I hope that I have managed to walk this intellectual tightrope without losing too many followers. At the very least, I hope to convince readers in a variety of fields that they are often talking about exactly the same physics as their colleagues in other subfields, albeit in a potentially different language. Also, I hope that readers gain the confidence to approach some of the higher-level texts on this topic that have appeared in recent years.

From a more personal point of view, this book summarises over a decade of my own research career and interests. These have been some of the most enjoyable years of my scientific life to date, enhanced enormously by the warm and friendly characters who populate this particular thicket of contemporary theoretical physics. When different theories come together, different people do too. That those people have been so keen in their intelligence, whilst also so welcoming in their humour and good spirit, has been a continual source of confidence and inspiration for me. Given the length of this book, I have naturally had to restrict myself to a certain range of topics that I personally felt would give a flavour of both the conceptual and practical aspects of the double copy. I mean no offence to those whose favourite topics have been omitted, and I should also stress that any misunderstandings, misrepresentations or similarly shameful signatures of general ignorance are entirely my own fault.

I am very grateful to Laurent Chaminade at World Scientific for initiating this project, also to Nambirajan Karuppiah, Ana Ovey and Shi Ying for their amazing help in the production process.

This book is dedicated to my twin brother Martin, who is himself a physicist. We have literally known each other since before we were born, and within that time have created many a jape, song and even a textbook on particle physics. To date, however, we have never managed to write a joint research paper. Who knows – perhaps this book shall finally pique his interest?

My second dedicatee is my long-suffering sister Nicola. Not only is she the funniest person I know, but she has also put up for years with my brother and I disrupting family gatherings with our incessant physics chat. She also reminds me that one cannot think or talk about physics all the time: it is equally important to play silly and rather violent games of pattycake, much to the amusement of our baffled children.

About the Author

Dr. Chris D. White is a Reader in Theoretical Physics at Queen Mary University of London. Originally from Cornwall, he studied at the University of Cambridge, obtaining a PhD on the structure of the proton, before holding positions in Amsterdam, Durham and Glasgow. Chris has published over 70 research papers on a wide range of topics in high-energy physics, ranging from practical calculations for particle accelerators such as the Large Hadron Collider, to relationships between particle physics and gravity. He has also won teaching awards recognising his efforts to make physics accessible to students from underrepresented backgrounds.

Contents

Chapter 1

Introduction

1.1 The Current State of Fundamental Physics

Perhaps more than any other time during the collective history of the human species, the last century and a half has seen an astonishing escalation in our understanding of the universe. In particular, the twin developments of relativity and quantum mechanics revolutionised how we think about the world around us to such a radical extent that the philosophical implications are still being settled. Questions of where the universe came from, how it will end, and how it operates at its most elemental fall under the remit of *fundamental physics*, whose basic ideas are fairly simple to state: the universe contains *matter*, which is acted on by *forces*. All forces that we observe turn out to be consequences of four *fundamental forces*, which at low energies appear as electromagnetism, the weak and strong nuclear forces, and gravity. The basic building blocks of matter are also known in terms of *fundamental particles*. These include the *leptons*, comprising the electron (usually written e^- for short) and heavier electron-like particles known as the muon (μ^-) and tauon (τ^-), as well as associated particles called *neutrinos* (ν_e, ν_μ, ν_τ). Then there are the *quarks* that live inside strongly interacting particles, such as the proton and neutron. These consist of six different types, or "flavours", which for historical reasons are called up (u), down (d), charm (c), strange (s), top (t) and bottom (b). We summarise the properties of these particles in Table 1.1, noting in particular that they form three *generations*. Quite why there are only three generations, and also why the fundamental particles have the enormous

1

Table 1.1. Fundamental matter particles and the three generations of particles, their masses, and the forces they feel.

		Quarks			
		Mass/GeV	Strong	Weak	EM
1st Generation	$\begin{pmatrix} u \\ d \end{pmatrix}$	0.002 0.005	✓ ✓	✓ ✓	✓ ✓
2nd Generation	$\begin{pmatrix} c \\ s \end{pmatrix}$	1.28 0.095	✓ ✓	✓ ✓	✓ ✓
3rd Generation	$\begin{pmatrix} t \\ b \end{pmatrix}$	173 4.18	✓ ✓	✓ ✓	✓ ✓

		Leptons			
		Mass/GeV	Strong	Weak	EM
1st Generation	$\begin{pmatrix} e^- \\ \nu_e \end{pmatrix}$	0.0005 ?	× ×	✓ ✓	✓ ×
2nd Generation	$\begin{pmatrix} \mu^- \\ \nu_\mu \end{pmatrix}$	0.106 ?	× ×	✓ ✓	✓ ×
3rd Generation	$\begin{pmatrix} \tau^- \\ \nu_\tau \end{pmatrix}$	1.78 ?	× ×	✓ ✓	✓ ×

Table 1.2. Fundamental forces and the particles that carry them.

Force	Carrier(s)	Mass/GeV
Electromagnetism	γ (photon)	0
Strong	g (gluon)	0
Weak	$W^\pm$, Z^0	80.39 (W bosons), 91.188 (Z boson)

range of masses that they have, remain mysterious. Apart from the particles listed in the table, there are also *antiparticles*, which are identical in mass to their corresponding matter particles, but with other quantum numbers (such as electromagnetic charge) reversed.

As we will see in the following, the forces turn out to be also carried by particles, whose properties we summarise in Table 1.2. We see that the electromagnetic and strong forces are each associated with a single massless particle, whereas the weak interaction is carried by three massive particles. The masses of the latter – and indeed the masses of the matter particles in Table 1.1 – cannot be considered arbitrarily, but must be generated by interaction with an additional particle called the *Higgs boson*. This is itself intimately related to the fact that the electromagnetic and weak forces, which appear separate at low energies, in fact mix with each other at higher energies to make a combined *electroweak theory*. We can then write a single theory that combines the electroweak and strong interactions, together with the matter particles themselves. This is known as the *standard model of particle physics*. It is a type of theory called a *quantum field theory* (QFT), in which framework one unavoidably ends up when describing particles subject to both special relativity (SR) and quantum mechanics (QM). To date, the SM has withstood every experimental test at collider physics experiments, such as the Large Hadron Collider (LHC) and its predecessors. But there are clear problems and puzzles in nature, which the SM does not address. Examples include why matter dominates over antimatter in the universe, why the Higgs boson has a low mass (when quantum corrections suggest its mass should be much higher), and what the nature of the mysterious *dark matter* and *dark energy* are, required to fit astrophysical observations. Moreover, the Higgs boson was only discovered as recently as 2012, such that the discovery of particle physics beyond the standard model could be just around the corner. General arguments dictate that any new physics theory must look like a QFT at sufficiently low energies, which allows us to recycle our knowledge of how to do calculations in the SM in order to probe possible new physics effects.

Gravity is conspicuously missing from almost all discussions involving particle physics. Although all massive particles interact gravitationally, this force is incredibly weak compared to the electromagnetic, weak and strong forces, and so is barely noticeable on subatomic scales. That we even notice gravity in our everyday lives is due to the fact that we live on a very massive planet, whose entire mass pulls us towards its core. Even then, we can easily overcome gravity if we want to and leap off the ground using our much smaller legs! Nevertheless, it is gravity that controls the dynamics of the largest objects in the universe, such as planets, stars, (clusters of) galaxies, and even the entire universe itself. Our best current theory of gravity is *general relativity* (GR) which, like the revolutionary theories mentioned above, has profound consequences for how we think about the world around us. Pre-GR, space and time were passive entities, providing merely the stage upon which our theories of physics played out. By contrast, GR stipulates that spacetime (an amalgam of space and time already necessitated by SR) itself becomes dynamical: like a sheet of rubber, it warps in response to the matter and energy it contains. Particles moving in such a curved space do not follow straight lines, but instead can appear to move towards each other, and it is this geometric effect that GR tells us represents the force of gravity. The equations of GR prescribe, in very precise terms, how the distribution of mass and energy throughout the universe governs the curvature of the universe at all points in spacetime. We will see later on how to write this in the appropriate mathematical language, but the consequences of these equations are dramatic. Specific solutions correspond, in principle, to the complete structure of spacetime i.e. to the history and future of the whole universe all at once. One such solution – that fits observations of our own universe rather well – corresponds to a universe that expands outwards from a finite time in the past, which has become well known as the *Big Bang Theory*. Other solutions represent specific objects localised in certain spatial regions, and we may take these to approximate actual objects in our universe, provided these solutions tend to empty space as we go far away. Some of these objects in GR correspond to regions of space where the curvature becomes so high that not even light can escape. These are *black holes*, and our ability to study these objects in real life has significantly advanced in recent years. A further interesting solution of note is that of *gravitational waves*. These correspond to ripples in the fabric of spacetime that travel at the speed of light. Returning to the above analogy of spacetime as a sheet of rubber, it is as if someone wobbles the sheet at one end, so that it oscillates, carrying energy away. Although indirect evidence for these wave solutions was found already in the 1970s, they were directly observed for the first time by the LIGO experiment in 2015. In order for us to be able to detect gravitational waves on Earth, they must be generated by very extreme events, such as the collision of black holes or neutron stars. Thus, their detection opens up a whole new way of observing the universe as well as providing motivation for developing new calculational techniques for GR.

Despite the myriad of successes that GR has given us, it remains a classical theory, which does not include the effects of quantum mechanics. Furthermore, it breaks down at extreme places in our universe, such as at the centre of black holes and the beginning of the Big Bang itself, due in both cases to the fact that the curvature of spacetime becomes infinite. It is widely believed that a quantum theory of gravity is needed in order to circumvent this problem, and it is perhaps also conceptually desirable that gravity be quantum, given that this is true of the other forces in nature. However, apart from some tantalising hints of what a quantum theory of gravity might look like (such as the phenomenon of *Hawking radiation* from black holes or the fact that the latter possess a certain measure of entropy), we have not been able to conclusively settle its precise details. This is not helped by the fact that gravity is such a weak force, which makes it very difficult to directly test experimentally.[1] But there are also theoretical reasons for this impasse: attempts to apply the same quantum field-theoretic techniques to gravity, as to the other forces, results in nonsense results and, ultimately, a breakdown of predictability. To understand more about why this is, we need to describe the basic ideas of QFT in more detail.

1.2 The Ideas of QFT

In our beginning university studies in physics, we typically learn the laws of classical mechanics laid down several centuries ago, which we may unambiguously call *Newtonian mechanics*. We then learn that fast-moving objects obey different mechanical laws, namely those of *special relativity*, which was first motivated historically by the fact that the equations of electromagnetism predict a constant light speed, independent of which inertial frame we might happen to be in. At small scales, the laws get replaced again, this time by *quantum mechanics*, and we may depict this set of laws schematically as in Table 1.3. It follows that there must be a fourth theory to fill in the table, which applies for very small objects that are moving very fast. This is the role played by quantum field theory, and it includes the effects of both SR and QM. Unsurprisingly, the resulting edifice

---

[1]There are important exceptions to this, such as the temperature fluctuations in the *cosmic microwave background* that fills the universe, which may carry information about quantum fluctuations in the early universe.

Table 1.3. Range of physical theories applying to different physical situations; all theories in the upper and left-hand panels emerge as limits of quantum field theory.

		Speed of Object	
		Slow	Fast
Size of Object	Big	Newtonian Mechanics	Special Relativity
	Small	Quantum Mechanics	Quantum Field Theory

ends up being much more complicated than either of the theories that preceded it, but some of the most basic ideas of QFT are encountered very early on in our studies and are actually very simple to state.

To this end, let us consider the first QFT that was studied historically, namely that representing electromagnetism. We know in classical physics that there are both electric and magnetic *fields*, namely mathematical objects that associate an electric or magnetic field vector (which may be zero) at each point in spacetime. Alternatively, we can use a formalism (reviewed in the following chapter) in which electric and magnetic effects are combined into a single *electromagnetic field*. In any given situation, the field can be found by solving certain *field equations*, describing how the field evolves in space and time, given various sources (charges and currents) that may be present. For electromagnetism, these are the *Maxwell equations*, and under particular circumstances, they can be reduced to the wave equation, solutions of which are the famous *electromagnetic waves*. If we include the effects of quantum mechanics – in other words, we transition from the classical field theory of electromagnetism to the quantum field theory – then electromagnetic waves of a given frequency ν no longer have a continuous range of energies. Rather, the energy is forced to be *quantised* in units of $h\nu$, where h is Planck's constant. It is as if the field arrives in discrete lumps or "quanta", and indeed, this hypothesis was pivotal in explaining various experimental results (such as the spectrum of blackbody radiation or the emission of electrons from irradiated metals) in the early days of QM. We call these quanta *photons*, and they are the particles associated with the electromagnetic field that we already saw in Table 1.2.

Remarkably, this story generalises and explains how all of the matter and force particles occur in nature. That is, *every* type of fundamental particle (be it matter or force) is associated with a field filling all space, where different types of fields are needed in general. These various fields are described by certain field equations in each case, such that all of these field equations end up having wave-like solutions, as in electromagnetism. Then, the effect of "quantising" the theory is to create the particles in Tables 1.1 and 1.2. We should not think of these as being the same as the particles we first encounter in Newtonian mechanics: what we mean by a "particle" in QFT is a very different animal, which carries with it some amount of wave-like behaviour due to its origin as a quantum of a wave-like solution of a field. But it is perhaps rather humbling to think that the "stuff" that we are made of and the forces that hold this stuff together are all ultimately consequences of oscillating fields.

Another consequence of QFT is the existence of antimatter, which we have already mentioned above. Once again, it is relatively simple to comprehend how this arises upon trying to combine SR and QM. First, let us note that in non-relativistic mechanics, the energy E of a particle is given in terms of its momentum $\boldsymbol{p}$ and mass m by

$$E = \frac{\boldsymbol{p}^2}{2m}, \tag{1.1}$$

such that we have $E > 0$ in general. In SR, on the other hand, we instead have the relation

$$E^2 = \boldsymbol{p}^2 c^2 + m^2 c^4 \Rightarrow E = \pm\sqrt{\boldsymbol{p}^2 c^2 + m^2 c^4}, \tag{1.2}$$

where now E and $\boldsymbol{p}$ represent the appropriate relativistic generalisations of energy and momentum. The energy can now take positive or negative values, and in the classical theory, this is not a problem: we can simply assume that all physical particles have positive energies, such that they will remain as such. In a quantum theory, however, particles can transition between states of definite energy, and thus, there would be nothing to stop particles cascading down to states of unbounded negative energy, leading to the release of infinite amounts of energy in the process! This is clearly unphysical, and indeed, QFT turns out to remedy this situation by successfully reinterpreting negative energy particles as positive energy *antiparticles*.

A novel interpretation of these is that we can think of them as being like conventional particles, but moving backwards in time. We will use this fact later, but to see where it comes from, note that the angular frequency of a quantum particle is given by the formula

$$\omega = \frac{E}{\hbar}, \quad \hbar = \frac{h}{2\pi},$$

so that the time-dependent part of a wave-like solution is

$$\sim e^{iEt/\hbar}.$$

Reversing the sign of E is equivalent to reversing the direction of time, hence the novel interpretation of antiparticles stated above.

Despite its many complications, there is a sense in which QFT is much more conceptually natural than the other theories in Table 1.3. In non-relativistic quantum mechanics, for example, matter particles are described very differently to forces. In QFT, however, everything is described by a single concept – the *field* – although it remains true that different types of fields are used for matter and force particles. As a particular example of a QFT, the SM achieves a huge amount of unification of almost all of the physics we have ever known about into a single theoretical framework. It is a type of quantum field theory known as a gauge theory due to certain mathematical symmetries that we will explore in more detail in Chapter 2. Why, though, can QFT not describe general relativity, our current best theory of gravity?

1.3   The Problem with Gravity

As we will review in the following chapter, GR is traditionally thought about within the mathematical language of *differential geometry*, but this is not the only way to formulate it. Indeed, the language of QFT can be applied to GR, given that it has wave-like solutions: the gravitational waves mentioned above. Quanta of these waves are hypothetically called *gravitons*, and there is a resulting field that we can write down. However, upon including interactions in the theory, it ends up not being very predictive in that an infinite number of interactions are needed to fully describe the theory.

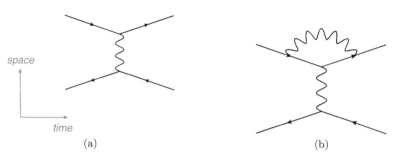

Fig. 1.1. (a) An electron and positron (solid lines) exchange a photon (wavy line); (b) a *one-loop process*, in which the first electron line emits and reabsorbs an additional photon.

To understand why this happens, let us first focus on the quantum theory of electromagnetism. Scattering processes in the theory can be represented pictorially using *Feynman diagrams* which, loosely speaking, are spacetime diagrams describing how the particles interact.[2] An example is shown in Figure 1.1(a), in which an electron and a positron (antielectron) interact by exchanging a photon, where the different particles are represented by different sorts of line. Note that the positron looks like an electron going backwards in time, in line with the comments above. A more complicated process is shown in Figure 1.1(b): now the incoming electron emits a photon, which is then reabsorbed by the outgoing electron in the final state. This creates two types of lines in the diagram: (i) external lines, which represent *real particles* that go to asymptotically large distances; (ii) internal lines representing the so-called *virtual particles* that are never observed directly and only exist for a short time. As we will see in Chapter 3, the rules of quantum field theory tell us that we have to sum over all possible energies/momenta of the virtual particles, and it turns out that this sum diverges. The origin of the problem can be traced to including arbitrarily high energies for the virtual photon, and such *ultraviolet (UV)* divergences plague

---

[2]Feynman diagrams are not to be taken too literally, as in the conventional formulation of QFT, each individual Feynman diagram actually represents a sum over all possible time orderings of the interaction vertices, so that special relativity is made manifest.

most relativistic quantum field theories. For large classes of theories, including those in the standard model of particle physics, we can remove the divergences by redefining the parameters entering the theory, such as interaction strengths and particle masses. This is called *renormalisation*, and theories that admit this procedure are called *renormalisable*. Gravity, on the other hand, is known to be *nonrenormalisable*: upon encountering UV divergences in GR, we have to add more and more interactions to the theory, whose accompanying parameters must be renormalised as we add more virtual particles. This is not actually a problem if we want to calculate quantum gravitational effects at low energies: we can simply add a sufficient number of potential interactions to the theory, and let experiment tell us what their coefficients are. This is called an *effective field theory*, and even renormalisable theories are nowadays considered to have potential non-renormalisable corrections. But the presence of unrenormalisable UV divergences is ultimately a signature that the QFT for gravity may break down at high energies and be replaced by a more fundamental underlying description. Noting that the uncertainty principle in quantum mechanics associates high energies or momenta with short distances in spacetime, this presumably has a bearing on why quantum corrections may be relevant at the centres of black holes or in the very early universe.

Various candidates exist for an underlying quantum theory of gravity. Some of these, such as loop quantum gravity or string theory, are qualitatively different *types* of theory. However, it may also be the case that there is a field theory of gravity, which reduces to GR at low enough energies and which does not suffer from UV divergences. One possibility is believed to be a theory known as $\mathcal{N} = 8$ supergravity, which contains much more symmetry than ordinary GR, in particular something called *supersymmetry*. This involves introducing fermionic or scalar fields, in addition to the graviton, whose properties are very precisely related to the graviton itself. The equations of such theories then remain the same if one interchanges the various fields in a prescribed way, which is the "symmetry" part of "supersymmetry". Different numbers of such extra fields can be introduced – referred to as having more than one supersymmetry. The index $\mathcal{N}$ then counts the number of distinct supersymmetries, and a higher value of $\mathcal{N}$ leads to a higher number of fields (and thus particles) in the theory. The case of $\mathcal{N} = 8$ turns out to be the highest amount of supersymmetry one can have in four-dimensional spacetime, and thus, this theory is sometimes also referred to as *maximal supergravity*. Research on the structure of UV divergences in this and other gravitational field theories was stalled for decades due to the stupendous complexity of the computations involved. In fact, this complexity is not limited to quantum aspects of gravity. Even if one only considers traditional classical GR, it remains true that exact solutions are hard to come by and approximate solutions become more and more difficult as the accuracy increases. Nevertheless, precise predictions are needed to get the most out of current and forthcoming gravitational wave experiments. Thus, it is imperative to develop new calculational methods in both classical and quantum gravity theories. These in turn will allow us to address practical, as well as conceptual, questions about the universe we live in.

1.4 The Double Copy

The aim of this book is to review a set of methods for addressing gravitational – and related – questions, which came to be known as the *double copy*. Inspired by previous work in string theory [1], the original incarnation of the double copy applied to quantities called *scattering amplitudes*, which are related to the probabilities for particles to interact in QFT. In particular, Refs. [2–4] argued that amplitudes in the gauge theories underlying forces such as those in the standard model can be straightforwardly recycled to produce scattering amplitudes in (quantum) gravity theories, where these make sense. In doing so, one must write the gauge theory results in a certain form, that suggests a powerful new relationship (*BCJ duality*) between the charges of fundamental particles and their kinematic degrees of freedom (e.g. momenta, polarisations). Then one must replace the charge information with an additional set of kinematic factors, hence the origin of the name *double copy*. Generation of scattering amplitudes via the double copy proves to be much more efficient than traditional calculations in quantum gravity and has allowed researchers to extend our knowledge of UV divergences – and their absence – in various supergravity theories [5–11]. However, unlike many of the developments in the theoretical physics of the past decades, the double copy also applies to theories without supersymmetry, including those that

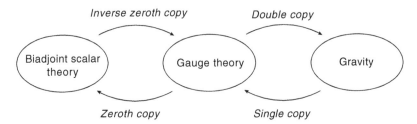

Fig. 1.2.   Various types of theory and the relationships between them.

are directly relevant to current experiments. Related correspondences apply to more exotic types of quantum field theory, which are not directly relevant to nature (and indeed can never be). If the double copy implies that we replace charge with kinematic information in scattering amplitudes, we can also go the other way and replace kinematic quantities with other types of charges. This is called the *zeroth copy* and leads to a theory known as *biadjoint scalar field theory*. We can summarise the different types of theories we have encountered, and the relationships between them, using the scheme of Figure 1.2. In fact, this ladder of theories is merely a subset of an ever-widening web of theories that have been found to obey similar correspondences (see e.g. Refs. [12–14]).

The double copy would be highly useful even if it applied merely to scattering amplitudes. However, intense research in the last decade has established that the scheme of Figure 1.2 is much broader than this. Quite how generally the double copy applies remains to be seen, but we now know that it extends to classical solutions of the various theories. That is, one may take the equations of motion corresponding to each of the theories in Figure 1.2 and find solutions for the fields that can be meaningfully mapped into each other by double-copy-like procedures. This is usually referred to as the *classical double copy* to distinguish it from the original incarnation for scattering amplitudes. However, the past few years have given us a much better understanding of why the classical double copy is a manifestation of the same correspondence that underlies the amplitude double copy, and indeed overlaps where relevant. In some cases, it is possible to take *exact* solutions in gravity theories and find counterparts living in a gauge or other theory. Known cases include some of the most famous solutions of general relativity, including (non-)spinning black holes. More generally, one cannot solve the equations of GR exactly for situations of astrophysical interest. However, one can instead solve them approximately using perturbation theory, and it is then known how to apply the double copy order by order. This allows us to take classical results in a (much simpler) gauge theory and turn them into gravitational quantities that would be much more difficult to calculate using traditional methods. Indeed, new results relevant for the study of gravitational waves have been obtained using this method [15–19]. The coming years are likely to see increased application of the double copy to astrophysics, allowing us to perform calculations that were previously impossible.

Apart from practically useful tools, the double copy suggests there is a profound structure underlying the field theories used in fundamental physics, which our traditional methods have obscured. One might even go so far as to suggest that the mathematical foundations of (quantum) field theory need to be rewritten in order to make this hidden structure manifest. This in turn offers to revolutionise our understanding of all of the theories in Figure 1.2, so that it is important to understand how generally the correspondences shown there apply. Work in recent years has continued to expand the network of theories that are interrelated by double-copy-like correspondences as well as the types of quantity that can be "copied". It has been intriguing to see how different physical effects, in very different theories, turn out to be related to each other in a very precise sense. What has been equally intriguing is the development of common languages that are able to simultaneously describe very different situations.

A problem of physics nowadays is its increasing specialisation, such that students start to diverge even at Masters level. The double copy opposes this tendency by potentially relating many subfields of physics e.g. high-energy physics, string theory, astronomy, cosmology and even condensed matter physics and optics [20]. However, it remains true that many research papers on this topic will remain incomprehensible to all but a few seasoned practitioners, albeit steadily increasing in number. The aim of this book will be to present many of the technical ideas of the double copy in such a way that they are accessible to a wider audience, who may then use this book as a springboard to higher-level review articles and research papers. We must begin by properly defining the theories that we will encounter throughout the rest of the book and that we have already seen in Figure 1.2. This is the subject of the following chapter.

Chapter 2

The Theories of Fundamental Physics

This chapter aims to review the theories that are discussed in the remaining chapters. As we have seen in the previous chapter, the double copy and related correspondences apply to a wide variety of theories both with and without supersymmetry. Our focus will be on more familiar theories that are also closer to practical applications in particle physics and gravity. We will necessarily have to assume much prior knowledge, given that a full introduction to each of the theories discussed here would fill an entire book! However, we will point the reader to more introductory references as necessary.

2.1 4-Vectors and Tensors

As is hopefully clear from the previous chapter, all of the theories we need to consider are *field theories*, and they also satisfy the requirements of special relativity. This in turn means that all of the relevant field equations can be written in terms of mathematical quantities that transform nicely under Lorentz transformations. These are 4-vectors, tensors and (for fermionic fields) spinors, and we will assume that the reader is already familiar with at least the first two. The following provides a rather swift recap: see e.g. Refs. [21, 22] for a much fuller and more introductory discussion.

We will work in *natural units* throughout, in which factors of c and $\hbar$ are ignored until further notice (or alternatively set to one). In the absence of gravity, our various field theories then live in *Minkowski spacetime*, such that a given spacetime position can be labelled by the 4-vector[1]

$$x^\mu = (t, \boldsymbol{x}) \equiv (x^0, x^1, x^2, x^3), \tag{2.1}$$

where $\boldsymbol{x}$ is the 3-vector spatial position and t the time coordinate. The Greek letter on the left-hand side is an index that ranges over $(0, 1, 2, 3)$ and picks out each component as required. Sometimes it will be convenient to label coordinates according to the notation of the right-hand side, and sometimes we will instead use the explicit spacetime coordinates (e.g. (t, x, y, z) in Cartesian coordinates). The former creates a potential confusion e.g. x^2 on the right-hand side of Eq. (2.1) is the third component (and second spatial component) of the position 4-vector and not the x component squared! Hopefully, meanings will always be clear from the context.

Particles move about in general, and the trajectory of a particle in spacetime will be given by some curve $x^\mu(\tau)$, where this denotes the position of the particle at a *proper time* τ, corresponding to the time measured in a reference frame in which the particle is momentarily at rest. Given this trajectory, we can form the 4-velocity and 4-momentum

$$u^\mu = \frac{dx^\mu(\tau)}{d\tau}, \quad p^\mu = mu^\mu, \tag{2.2}$$

respectively. Here m is the rest mass, namely the mass of the particle that would be measured in the particle's rest frame. In a specific frame with spacetime coordinates $(t, \boldsymbol{x})$, it can be convenient to recall that proper time differences get dilated according to

$$dt = \gamma d\tau, \quad \gamma = \sqrt{1 - \boldsymbol{v}^2}, \tag{2.3}$$

where $\boldsymbol{v}$ is the conventional 3-velocity of the particle. Then the 4-momentum assumes the form

$$p^\mu = (m\gamma, m\gamma\boldsymbol{v}), \quad \boldsymbol{v} = \frac{d\boldsymbol{x}}{dt}. \tag{2.4}$$

This can be written in terms of the known relativistic energy E and momentum $\boldsymbol{p}$:

$$p^\mu = (E, \boldsymbol{p}), \tag{2.5}$$

---

[1]Note the use of natural units already in Eq. (2.1). In SI units, there would be an explicit factor of c in the time component.

satisfying the *energy–momentum relation*

$$E^2 - \boldsymbol{p}^2 = m^2. \tag{2.6}$$

The components V^μ are called the *contravariant* components of a 4-vector. We may also define the *covariant* components, according to the prescription (in Cartesian components)

$$V_\mu \equiv (V_0, V_1, V_2, V_3) = (-V^0, V^1, V^2, V^3). \tag{2.7}$$

That is, vectors with a downstairs index have their time-like component reversed with respect to their counterparts with an upstairs index. We may then define a dot product for 4-vectors:

$$V \cdot W = \sum_\mu V^\mu W_\mu = -(V^0)(W^0) + (V^1)(W^1) + (V^2)(W^2)$$

$$+ (V^3)(W^3). \tag{2.8}$$

This involves combining an upper index with a lower index and summing over all values of the index, a process known as *contraction*. We will see a great many index contractions in this book and thus will follow Einstein's marvellous *summation convention*, in which any repeated index is assumed to be summed over unless otherwise stated, and all explicit summation signs suppressed. Furthermore, we should note that different conventions exist for 4-vectors across the literature. Other people may, for example, label indices as going from 1 to 4, and where either the first or fourth component is taken to be the time-like component. Even if following the same index convention as this book, they might choose to define the dot product as having a plus sign associated with the zeroth components and a minus sign for the others. We are thus using what is known as the *mostly plus* convention, as opposed to *mostly minus*.

The above dot product is defined only in Cartesian coordinates. But we may generalise this by introducing the *metric tensor* and its inverse

$$\eta_{\mu\nu} = \begin{pmatrix} -1 & 0 & 0 & 0 \\ 0 & 1 & 0 & 0 \\ 0 & 0 & 1 & 0 \\ 0 & 0 & 0 & 1 \end{pmatrix}, \quad \eta^{\mu\nu} = \begin{pmatrix} -1 & 0 & 0 & 0 \\ 0 & 1 & 0 & 0 \\ 0 & 0 & 1 & 0 \\ 0 & 0 & 0 & 1 \end{pmatrix}, \tag{2.9}$$

respectively. Then one may easily verify the raising and lowering properties

$$V_\mu = \eta_{\mu\nu} V^\nu, \quad V^\mu = \eta^{\mu\nu} V_\nu, \tag{2.10}$$

as well as the fact that Eq. (2.8) may be written as

$$V \cdot W = \begin{pmatrix} V^0 & V^1 & V^2 & V^3 \end{pmatrix} \begin{pmatrix} -1 & 0 & 0 & 0 \\ 0 & 1 & 0 & 0 \\ 0 & 0 & 1 & 0 \\ 0 & 0 & 0 & 1 \end{pmatrix} \begin{pmatrix} W^0 \\ W^1 \\ W^2 \\ W^3 \end{pmatrix} = \eta_{\mu\nu} V^\mu W^\nu. \tag{2.11}$$

Here we give both the matrix interpretation of this formula and the index form. By analogy with 3-vector algebra, we can associate a squared magnitude with every 4-vector, given by

$$V^2 = V \cdot V. \tag{2.12}$$

Then 4-vectors with $V^2 < 0$, $V^2 = 0$, and $V^2 > 0$ are called *time-like*, *null*, and *space-like*, respectively. The metric tensor gets its name from the fact that it tells us how to measure distances in a given space. Consider, for example, zooming in on a point with coordinates x^μ and then considering the spacetime "distance" to point $x^\mu + dx^\mu$. The squared spacetime "length" associated with the small displacement is, from the above magnitude,

$$ds^2 = \eta_{\mu\nu} dx^\mu dx^\nu = -(dt)^2 + (dx)^2 + (dy)^2 + (dz)^2, \tag{2.13}$$

which acts like a kind of four-dimensional generalisation of Pythagoras' theorem, albeit where time and space components are counted with opposite signs. Equation (2.13) in turn allows us to see how the components of $\eta_{\mu\nu}$ change if we move to a non-Cartesian coordinate system. Consider transforming to some new coordinates $\{x'^\mu\}$, defined as functions of the old coordinates $\{x^\mu\}$. The chain rule tells us

$$dx^\mu = \frac{\partial x^\mu}{\partial x'^\alpha} dx'^\alpha, \tag{2.14}$$

such that we may write

$$ds^2 = \eta_{\mu\nu} \frac{\partial x^\mu}{\partial x'^\alpha} \frac{\partial x^\nu}{\partial x'^\beta} dx'^\alpha dx'^\beta \tag{2.15}$$

$$= \eta'_{\alpha\beta} dx'^\alpha dx'^\beta. \tag{2.16}$$

By comparing Eqs. (2.13) and (2.16), we find that the metric tensor in the new coordinates is given by

$$\eta'_{\alpha\beta} = \eta_{\mu\nu} \frac{\partial x^\mu}{\partial x'^\alpha} \frac{\partial x^\nu}{\partial x'^\beta}. \tag{2.17}$$

Likewise, the inverse metric tensor is found to transform as

$$\eta'^{\alpha\beta} = \eta^{\mu\nu} \frac{\partial x'^\alpha}{\partial x^\mu} \frac{\partial x'^\beta}{\partial x^\nu}. \tag{2.18}$$

Throughout this book, we will frequently make use of the 4-vector operator

$$\partial_\mu \equiv \frac{\partial}{\partial x^\mu}, \tag{2.19}$$

which in Cartesian coordinates implies

$$\partial_\mu = \left(\frac{\partial}{\partial t}, \nabla \right), \quad \partial^\mu = \left(-\frac{\partial}{\partial t}, \nabla \right), \tag{2.20}$$

where ∇ is the usual 3-vector gradient operator. To transform Eq. (2.19) to another coordinate system, $\{x'^\mu\}$, we may use the chain rule to write

$$\partial'_\mu = \frac{\partial}{\partial x'_\mu} = \frac{\partial x^\alpha}{\partial x'_\mu} \frac{\partial}{\partial x_\alpha} = \frac{\partial x^\alpha}{\partial x'_\mu} \partial_\alpha. \tag{2.21}$$

Equations (2.17), (2.18), and (2.21) are special cases of a general transformation law for objects with multiple indices, known collectively as *tensors*. Consider the quantity

$$V^{\alpha_1\alpha_2...\alpha_n}_{\beta_1\beta_2...\beta_m}(dx^{\beta_1} \ldots dx^{\beta_m})(dx_{\alpha_1} \ldots dx_{\alpha_n}), \tag{2.22}$$

involving some mixed-index tensor components $V^{\alpha_1\alpha_2...\alpha_n}_{\beta_1\beta_2...\beta_m}$. Similar arguments to those leading to Eq. (2.17) can be used to derive the following law for transforming to a new (primed) coordinate system:

$$V'^{\alpha_1\alpha_2...\alpha_n}_{\beta_1\beta_2...\beta_m} = V^{\mu_1...\mu_n}_{\nu_1...\nu_m} \frac{\partial x'^{\alpha_1}}{\partial x^{\mu_1}} \cdots \frac{\partial x'^{\alpha_n}}{\partial x^{\mu_n}} \frac{\partial x^{\nu_1}}{\partial x'^{\beta_1}} \cdots \frac{\partial x^{\nu_m}}{\partial x'^{\beta_m}}. \tag{2.23}$$

All of the theories that we are going to see can be defined by equations of motion for the fields involved, where all such equations will be expressible in terms of 4-vectors and tensors. The convenience of this language is that any such field equation will have the same form in different spacetime coordinate systems, which is ultimately a consequence of the transformation law of Eq. (2.23).

2.2 Electromagnetism

Having introduced the crucial language of tensors in the previous section, let us now consider our first theory. Arguably, the simplest one to start with is *electromagnetism*. As described in the previous chapter, this is one of the four fundamental forces in nature, where the other forces in the standard model can be viewed as more complicated generalisations of it. Furthermore, electromagnetism is familiar to us from our beginning undergraduate studies (if not before), although perhaps not in the relativistic formulation presented here. If the following is unfamiliar, the reader is advised to consult Ref. [22] for a comprehensive introduction.

2.2.1 *Covariant formulation*

As the name suggests, electromagnetism unifies two well-known phenomena in nature, namely electricity and magnetism. In our first exposure to the theory, we typically see these represented by separate *electric* and *magnetic fields*, commonly denoted by $\boldsymbol{E}$ and $\boldsymbol{B}$, respectively. Another way to formulate the theory is to introduce the *electrostatic potential* V and the *magnetic vector potential* $\boldsymbol{A}$, such that the physical fields are given by

$$\boldsymbol{E} = -\nabla V - \frac{\partial \boldsymbol{A}}{\partial t}, \quad \boldsymbol{B} = \nabla \times \boldsymbol{A}. \tag{2.24}$$

The potentials may be combined to form the *potential 4-vector*

$$A^\mu = (V, \boldsymbol{A}), \tag{2.25}$$

and one may then define the *field strength tensor*

$$F_{\mu\nu} = \partial_\mu A_\nu - \partial_\nu A_\mu. \tag{2.26}$$

Using Eqs. (2.20) and (2.25), the components of this tensor in Cartesian coordinates are found to be

$$F_{\mu\nu} = \begin{pmatrix} 0 & E_x & E_y & E_z \\ -E_x & 0 & -B_z & B_y \\ -E_y & B_z & 0 & -B_x \\ -E_z & -B_y & B_x & 0 \end{pmatrix}. \tag{2.27}$$

That is, the components of the physical electric and magnetic fields populate the field strength tensor, such that the latter becomes a natural object for unifying both effects into a single mathematical description. We may note that the field strength tensor is anti-symmetric ($F_{\mu\nu} = -F_{\nu\mu}$), which follows straightforwardly from the definition of Eq. (2.26). A general antisymmetric tensor in four dimensions has six independent components, which explains why we have to use such an object to include the two 3-vector fields $\boldsymbol{E}$ and $\boldsymbol{B}$. Armed with the field strength tensor, we may write the defining equations of electromagnetism, which are usually referred to as the *(covariant) Maxwell equations*.[2] The first equation describes how electric and magnetic fields are generated by charged particles, where the latter may be stationary or moving in any given coordinate frame. Moving charges are more commonly called *currents*, and we can talk about the distribution of charges and currents by introducing the *charge density* $\rho(\boldsymbol{x})$ at a given point in space and the 3-vector *current density* $\boldsymbol{J}(\boldsymbol{x})$. These may be combined into a single *4-vector current density*

$$j^\mu = (\rho, \boldsymbol{J}) \tag{2.28}$$

such that conservation of charge is expressed by the *continuity equation*

$$\partial_\mu j^\mu = \frac{\partial \rho}{\partial t} + \nabla \cdot \boldsymbol{J} = 0. \tag{2.29}$$

Then our first Maxwell equation is

$$\partial^\nu F_{\mu\nu} = j_\mu. \tag{2.30}$$

---

[2]The word "covariant" here means "transforms nicely under Lorentz transformations" and distinguishes the tensor form of the Maxwell equations from their usual 3-vector expression that we see in what follows.

This has the current on the right-hand side and the electric and magnetic fields on the left-hand side. Thus, it indeed tells us that the former produces the latter.

The second Maxwell equation is usually written as

$$\partial_\mu F_{\nu\alpha} + \partial_\alpha F_{\mu\nu} + \partial_\nu F_{\alpha\mu} = 0 \tag{2.31}$$

and is also known as the *Bianchi identity*. In fact, this follows straightforwardly from the definition of Eq. (2.26): upon differentiating and forming the above combination, the various terms cancel out due to the fact that the order in which partial derivatives are taken does not matter. The explicit form of Eq. (2.27) can be used to show that Eqs. (2.30) and (2.31) are equivalent to the four 3-vector equations[3]

$$\nabla \cdot \boldsymbol{E} = \rho,$$

$$\nabla \cdot \boldsymbol{B} = 0,$$

$$\nabla \times \boldsymbol{E} = -\frac{\partial \boldsymbol{B}}{\partial t},$$

$$\nabla \times \boldsymbol{B} = \boldsymbol{J} + \frac{\partial \boldsymbol{E}}{\partial t}, \tag{2.32}$$

which are familiar to us from a younger age!

So much for the fields. To complete the theory, we also need to say how charged particles respond to the electromagnetic field. Let us take a single particle with charge q and 4-velocity (4-momentum) u^μ (p^μ). Then the *Lorentz force equation* states that

$$\frac{dp_\mu}{d\tau} = q F_{\mu\nu} u^\nu. \tag{2.33}$$

Note the use of the proper time τ of the particle on the left-hand side. This ensures that the left-hand side is a 4-vector, which it had better be given that the right-hand side is composed of tensor quantities, with one free-index, and thus is also a 4-vector. To interpret

---

[3]One would normally see various factors of the permittivity ϵ_0 and permeability μ_0 of free space on the right-hand side of Eq. (2.32). However, our choice of natural units and the fact that $c = 1/\sqrt{\mu_0\epsilon_0}$ means that we are effectively ignoring all factors of μ_0 and ϵ_0.

Eq. (2.33), note that the left-hand side contains a rate of change of 4-momentum, which is a kind of relativistic generalisation of a force in Newtonian mechanics. The right-hand side then says that this "force" is linear in the electromagnetic field and in the velocity of the particle. Equation (2.33) implies the 3-vector equation

$$\boldsymbol{F} = q(\boldsymbol{E} + \boldsymbol{v} \times \boldsymbol{B}), \tag{2.34}$$

where $\boldsymbol{F}$ is the 3-vector force acting on the particle and $\boldsymbol{v}$ its velocity. Equation (2.34) is how we usually first encounter the Lorentz force, alongside the 3-vector Maxwell equations of Eqs. (2.32).

The theory of electromagnetism has a huge number of practical applications by itself but is also pivotal in understanding the other forces in nature. There is an elegant mathematical framework underlying the theory that allows it to be generalised to more complicated forces. We explore this in more detail in the following section.

2.2.2   *Electromagnetism as a gauge theory*

When we move from the 3-vector form of the Maxwell equations of Eq. (2.32) to the covariant form (Eqs. (2.30) and (2.31)), the information content is the same. However, the number of equations reduces, and thus it starts to look like there is some sort of powerful organising principle behind why the equations *necessarily* have to take such a form. Indeed, there is such a principle, and it involves a rather abstract symmetry of the theory called *gauge invariance*. Perhaps the best way to introduce this is to ignore electromagnetism for the moment and to consider the equations describing the matter particles that we see in nature. Arguably, the most familiar charged particle we encounter is the *electron*, and according to the ideas described in the previous chapter, it will be described by a field $\Psi(x)$ filling all of spacetime, such that quanta of this field correspond to individual particles. The relevant field equation for a free (non-interacting) electron field is the *Dirac equation*[4]

$$(\gamma^\mu \partial_\mu + m)\Psi(x) = 0, \tag{2.35}$$

---

[4]Equation (2.35) will look different to many reference books on quantum field theory, due to our use of the mostly plus metric. In the mostly minus metric,

where γ^μ is a 4-vector of certain constants and m the mass of the electron. If you have seen this equation before, you will know that $\Psi(x)$ is actually a 4-component object called a *spinor*, which must consequently transform in a certain special way under Lorentz transformations. Likewise, the constants $\{\gamma^\mu\}$ then turn out to be 4×4 matrices (known as *Dirac matrices*), and there is also an identity matrix in the second term for consistency. None of this actually matters for the following discussion, and you may simply take Eq. (2.35) on trust if you have not seen it before.[5]

Note that $\Psi(x)$ is a complex field in general, and thus it may carry a *phase* at each point in space. Given that phases can only go from 0 to 2π, we can represent the phase of the field at each point on a unit circle, as shown in Figure 2.1(a). In order to specify the phase, we must specify a particular point on the circle as constituting "zero phase", which is represented by the dotted line, and then a convention for the direction of phase e.g. anticlockwise. Of course, where we specify the zero of phase is a human convention and not a property of nature. All physically measurable quantities involving electrons

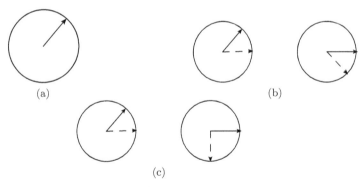

(a)   (b)

(c)

Fig. 2.1. (a) The phase of the electron field at a given spacetime point can be represented as an arrow on a unit circle; (b) a global gauge transformation consists of changing the phase of the field by the same amount at all spacetime points (from the solid to the dashed arrows); (c) in a local gauge transformation, the phase can change by *different* amounts at different spacetime points.

---

there is an explicit factor of i in the derivative term and a minus rather than a plus sign in front of the mass m.

[5] An introductory exposition of the Dirac equation, and a fuller discussion of the ideas of this section, may be found in Ref. [22].

indeed turn out to be insensitive to this choice when calculated, and it implies that the theory has a certain symmetry: we can redefine what we mean by the zero of phase on the unit circle, provided we do this in the same way at all points in spacetime so that *phase differences* of the field at different points stay the same. An equivalent way of saying this is that we can keep the zero of phase the same at all points but instead rotate the arrows representing the phase at different points, by the same fixed amount. This procedure is shown in Figure 2.1(b) and is known as a *global gauge transformation*. The word "gauge" relates to the fact that we are recalibrating what we mean by phase, and the word "global" implies that we are doing this simultaneously at all points in spacetime.

Let us see how to confirm the earlier remarks mathematically. First, note that changing the phase of the electron field everywhere in spacetime amounts to multiplying it by a unit complex number:

$$\Psi(x) \rightarrow e^{i\alpha}\Psi(x). \tag{2.36}$$

Substituting this into the Dirac equation of Eq. (2.35), this becomes

$$(\gamma^\mu \partial_\mu + m)[e^{i\alpha}\Psi(x)] = e^{i\alpha}(\gamma^\mu \partial_\mu + m)\Psi(x) = 0, \tag{2.37}$$

where the second equality follows from Eq. (2.35). Thus, the Dirac equation is indeed preserved under global gauge transformations, as expected.

In fact, the actual theory obeyed by electrons in our universe turns out to have infinitely more symmetry than this. Without asking why for the moment, let us now try to construct a theory with a *local gauge symmetry*, namely one in which we can change the phase of the field by different amounts at different points in space, as shown in Figure 2.1(c). This amounts to letting the arbitrary phase parameter α in Eq. (2.36) become a function of spacetime position $\alpha(x)$, and the Dirac equation involving the new field becomes

$$(\gamma^\mu \partial_\mu + m)[e^{i\alpha(x)}\Psi(x)] = e^{i\alpha(x)}[\gamma^\mu \partial_\mu + m + i\gamma^\mu \partial_\mu \alpha(x)]\Psi(x). \tag{2.38}$$

We see, unsurprisingly, that the Dirac equation is not invariant under this transformation: an extra term appears involving the arbitrary function $\alpha(x)$, and the origin of this term is the fact that the derivative operator ∂_μ acts both on the original field $\Psi(x)$ and the function $\alpha(x)$. We can, however, insist upon local gauge invariance if we wish, by replacing the derivative ∂_μ with a so-called *covariant derivative* D_μ, such that

$$D_\mu \Psi(x) \to e^{i\alpha(x)} D_\mu \Psi(x) \tag{2.39}$$

under local gauge transformations. Then the modified Dirac equation

$$(\gamma^\mu D_\mu + m)\Psi(x) = 0 \tag{2.40}$$

is indeed locally gauge-invariant, provided we can find a suitable D_μ. To do this, let us guess the following form:

$$D_\mu = \partial_\mu + ieA_\mu(x), \tag{2.41}$$

where $A_\mu(x)$ is a 4-vector that depends on spacetime position in general, and the factors in the second term are conventional. If Eq. (2.41) doesn't work, then we can try something else, but in order to test it, we need to know how the quantity A_μ itself transforms under local gauge transformations. Let us parametrise our ignorance by writing that Eq. (2.36) implies

$$A_\mu \to A_\mu + \delta A_\mu, \tag{2.42}$$

for some change δA_μ, that may itself depend on the gauge transformation function $\alpha(x)$. Then Eqs. (2.41) and (2.42) imply

$$D_\mu \Psi(x) \to e^{i\alpha(x)}[D_\mu + i\partial_\mu\alpha(x) + ie\delta A_\mu]\Psi(x). \tag{2.43}$$

We see that we can indeed enforce Eq. (2.39), provided we take

$$\delta A_\mu = -\frac{1}{e}\partial_\mu\alpha(x), \tag{2.44}$$

which is clearly allowed. It follows that we can construct a locally gauge-invariant Dirac equation, by introducing an extra field $A_\mu(x)$, that transforms in tandem with the electron field. But how are we to interpret such a field? There are clearly infinitely many arbitrary functions $\alpha(x)$ that we can consider and thus infinitely many potential fields A_μ that we can construct. This arbitrariness means that Eq. (2.40) cannot be a sensible physical theory by itself.

To complete the theory, we can take the bold step of insisting that $A_\mu(x)$ is a genuine physical field in nature (or, at least, that it is related to physically measurable things!). If so, there must then be a further equation of motion for $A_\mu(x)$, such that combining this with Eq. (2.40) should tell us, in any given situation, what the electron and A_μ fields are doing. Typically, equations of motion for fields contain derivatives of the field, and we must also demand local gauge invariance given that this was the whole motivation for introducing $A_\mu(x)$ in the first place. Note that the basic field derivative

$$\partial_\mu A_\nu(x) \to \partial_\mu A_\nu(x) - \frac{1}{e}\partial_\mu\partial_\nu\alpha(x) \tag{2.45}$$

is not gauge-invariant, due to the second term on the right-hand side. However, the antisymmetric combination

$$F_{\mu\nu} \to \partial_\mu A_\nu - \partial_\nu A_\mu \tag{2.46}$$

satisfies $F_{\mu\nu} \to F_{\mu\nu}$, and thus is indeed gauge-invariant! If we want a field equation that is linear in the field, and second-order in derivatives, we are led almost unavoidably to the combination

$$\partial^\nu F_{\mu\nu} = j_\mu, \tag{2.47}$$

for some 4-vector $j_\mu(x)$. Furthermore, the definition of Eq. (2.46) implies the Bianchi identity of Eq. (2.31). Comparing Eq. (2.47) with Eq. (2.30), we see that the field $A_\mu(x)$ satisfies the Maxwell equations. In other words, demanding that the electron field obey local gauge invariance *automatically* leads to electromagnetism! It is truly humbling to realise that the vast range of phenomena we experience in our everyday lives can be traced back to a single powerful symmetry principle. We can also now interpret the constant e appearing in Eq. (2.41). Substituting the explicit form of the covariant derivative into the modified Dirac equation of Eq. (2.40), this becomes

$$(\gamma^\mu\partial_\mu + m + ie\gamma^\mu A_\mu)\Psi(x) = 0. \tag{2.48}$$

The third term in this equation shows that the electromagnetic field couples to the electron field, which is entirely what we expect! It also tells us that we can interpret the constant e as the strength of this interaction, namely the magnitude of the charge of the electron.

The above remarks, whilst compelling, do not go so far as to explain where electromagnetism comes from: we have merely shifted the question to explaining where local gauge invariance comes from. Indeed, recent research work has attempted to remove the need for local gauge invariance in finding general arguments for the existence of various theories. It is also true that gauge invariance by itself does not completely fix the form of Eq. (2.30): there is an extra gauge-invariant term we could have added to the field strength tensor of Eq. (2.46) that turns out to correspond to the presence of magnetic monopoles, which are absent in Maxwell's theory. We see magnetic monopoles later on.

What we can say with confidence is that the principle of local gauge invariance plays a highly effective role in constraining the form of Maxwell's equations. This in turn suggests that similar symmetry principles may play a role in describing the other forces in nature. Indeed they do, as we explore next.

2.3 QCD and Yang–Mills Theory

As our next example of a fundamental force, let us take the *strong force* that binds quarks together to make hadrons, such as the proton and neutron. Quarks are spin-$1/2$ particles like the electron, and thus obey the Dirac equation. However, an additional complication is that quarks carry a type of charge called *colour*, which is distinct from the electromagnetic charge that they also carry. Unlike electromagnetic charge, colour charge comes in three separate types, which are usually referred to as *red, green,* and *blue*, or (r, g, b) for short. It should be stressed that these labels have nothing to do with actual colours – which are an electromagnetic property! Rather, they are simply historical labels that have withstood the test of time.

Given the above observations, we can use a vector of fields $\Psi_i(x) = (\Psi_r(x), \Psi_g(x), \Psi_b(x))$ to describe quarks, where the different components of the field tell us how much redness, greenness, and blueness there is at a given point in spacetime. Geometrically, we can think of an abstract *colour space* at each point in spacetime and represent the colours of the quark fields by an arrow in this space. This is shown in Figure 2.2, which is somewhat reminiscent of the phase of the electron field that we drew in Figure 2.1(a). In both cases, we

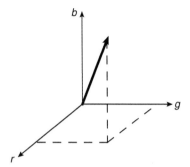

Fig. 2.2. The colour charges of the quark field at a given point in spacetime can be represented by an arrow in an abstract *colour space.*

have an abstract arrow associated with a field (in the present case, the vector of quark fields $\Psi_i(x)$). We saw that imposing symmetries under rotating the arrows in Figure 2.1 led to an additional field obeying Maxwell's equations of electromagnetism. Might something similar be true for the arrow in Figure 2.2? The answer of course is yes, but the route to the answer is much more long-winded.

First, let us ask what kind of symmetry we are talking about in the quark case. For electrons, we talked about the ability to redefine the zero of phase at all points in spacetime. The equivalent for quarks is that we can redefine what we mean by redness, greenness, or blueness (these labels are ultimately arbitrary) and thus can certainly perform global rotations of the colour arrows at each point in spacetime. Whereas global electron phase changes could be described by multiplying the field by a unit complex number, rotations of the colour arrow are clearly more complicated. They correspond to multiplying the quark field – which is vector-valued in colour space – with some sort of matrix:

$$\Psi(x) \to \mathbf{U}\Psi(x), \tag{2.49}$$

where the matrix $\mathbf{U}$ must be constant for a global transformation. Given the complex nature of the field $\Psi(x)$, $\mathbf{U}$ is a complex rotation matrix in general, and there is in fact a general way to characterise such matrices. First, total conservation of colour charge implies that the length of the colour arrow at each point in spacetime must be preserved by the rotation. If $\Psi^\dagger$ is the Hermitian conjugate of Ψ, we thus have

$$\boldsymbol{\Psi}^\dagger \boldsymbol{\Psi} \to \boldsymbol{\Psi}^\dagger \mathbf{U}^\dagger \mathbf{U} \boldsymbol{\Psi} = \boldsymbol{\Psi}^\dagger \boldsymbol{\Psi}. \tag{2.50}$$

Given $\mathbf{U}$ is an arbitrary complex rotation matrix, this implies the dual conditions

$$\mathbf{U}^\dagger \mathbf{U} = \mathbf{U} \mathbf{U}^\dagger = \mathbf{I}, \tag{2.51}$$

where $\mathbf{I}$ is the identity matrix in colour space. In words, $\mathbf{U}$ is *unitary*, and this is not the only restriction. Given that only proper rotations and not reflections of the colour arrow are being considered, the determinant of $\mathbf{U}$ must be one so that $\mathbf{U}$ is a so-called *special unitary matrix*. In our particular case, it has dimension three, due to the three quark colours.

So much for global gauge transformations of the quark field. It should hopefully also be clear what constitutes a local gauge transformation in this case, namely that the rotation of the colour arrow associated with the quark field can be *different* at different spacetime points:

$$\boldsymbol{\Psi}(x) \to \mathbf{U}(x) \boldsymbol{\Psi}(x). \tag{2.52}$$

Upon forming the Dirac equation for the transformed field, we find

$$(\gamma^\mu \partial_\mu + m)[\mathbf{U}\boldsymbol{\Psi}] = \mathbf{U}[\gamma^\mu \partial_\mu + m + \gamma^\mu \mathbf{U}^{-1}(\partial_\mu \mathbf{U})]\boldsymbol{\Psi}, \tag{2.53}$$

which is not the same as the Dirac equation for the untransformed field $\boldsymbol{\Psi}$, due to the third term in the brackets. As for the electromagnetic case, we can devise a *covariant derivative* that, when acting on the field $\boldsymbol{\Psi}$, behaves in the same way that the field does under local gauge transformations. It will, however, be more complicated than the electromagnetic case. Our field is now vector-valued (in colour space), and thus the covariant derivative is a matrix in general, where the desired transformation property is

$$\mathbf{D}_\mu \boldsymbol{\Psi} \to \mathbf{U} \mathbf{D}_\mu \boldsymbol{\Psi}. \tag{2.54}$$

We may again make an ansatz similar to Eq. (2.41), but where the matrix-valued nature of $\mathbf{D}_\mu$ in the present case means that we must generalise this to

$$\mathbf{D}_\mu = \mathbf{I}\partial_\mu + ig\mathbf{A}_\mu. \tag{2.55}$$

This has an identity matrix in the first term and a matrix-valued field (in colour space) $\mathbf{A}_\mu(x)$ in the second term, dressed by some conventional factors. Given that the identity matrix will simply multiply any other matrices without changing them, we will omit explicit factors of $\mathbf{I}$ in what follows, as is common throughout the literature. Furthermore, we will consider that the $\mathbf{A}_\mu$ gets modified by local gauge transformations to some new field $\mathbf{A}'_\mu$. Equations (2.54) and (2.55) then imply

$$\mathbf{U}(\partial_\mu + ig\mathbf{A}_\mu)\Psi = (\partial_\mu + ig\mathbf{A}'_\mu)\mathbf{U}\Psi. \qquad (2.56)$$

In words, gauge transforming the covariant derivative of the quark field according to Eq. (2.54) amounts to acting on the gauge-transformed field $\mathbf{U}\Psi$ with the gauge-transformed covariant derivative, where the latter consists of replacing $\mathbf{A}_\mu$ with $\mathbf{A}'_\mu$. Rearranging and using the fact that the field Ψ can be any physically allowed quark field, we find that Eq. (2.56) implies the following condition:

$$\mathbf{A}'_\mu = \mathbf{U}\mathbf{A}_\mu\mathbf{U}^{-1} + \frac{i}{g}(\partial_\mu\mathbf{U})\mathbf{U}^{-1}. \qquad (2.57)$$

That is, we can indeed construct a covariant derivative transforming according to Eq. (2.54), provided that the field $\mathbf{A}_\mu$ transforms according to Eq. (2.57). This is more complicated than Eqs. (2.42) and (2.44), but that is perhaps not surprising. In the electromagnetic case, local gauge transformations were simple rotations around a circle. Here, they are three-dimensional complex rotations! Armed with this covariant derivative, we can construct a locally gauge-invariant Dirac equation for the quark field by simply replacing the partial derivative with the covariant derivative:

$$(\gamma^\mu \mathbf{D}_\mu + m)\Psi = 0. \qquad (2.58)$$

Our next task will be to construct a field strength tensor for the field $\mathbf{A}_\mu$, that will turn it into a real physical field, analogous to how A_μ above ended up coinciding with the electromagnetic field. Indeed, the field $\mathbf{A}_\mu$ will turn out to correspond with the *gluon field*, responsible for binding quarks together to make hadrons. In order to arrive at the correct expression for the field strength, however, we need to take a detour to introduce some relevant mathematics.

2.3.1   *Lie groups and Lie algebras*

Sets of transformations form mathematical structures called *groups*. To define a group, we need a set of transformations, which we call the *elements* of the group. Then we need a rule for combining the elements a and b, where the notation ab means "first do the transformation b, then a". To form a well-defined group, the following properties must be true:

(i) The combination of any two group elements gives another element of the group.

(ii) The rule for combining any two elements a and b must be *associative*, meaning that

$$(ab)c = a(bc).$$

(iii) There must be an *identity element* e, corresponding to the transformation that does nothing. When combined with an arbitrary element a, one finds

$$ea = ae,$$

which gives another way to define e.

(iv) Each element a must have an *inverse element* a^{-1}, such that

$$aa^{-1} = a^{-1}a = e.$$

If you have not seen group theory before, the earlier rules will look very abstract indeed. However, they are certainly all satisfied by the local phase transformations we described earlier. First consider the electromagnetic case. There, our transformations consisted of rotations of an arrow on a unit circle, and the combination rule is simply "do the first rotation, then the second". Clearly, the combination of any two rotations is itself a rotation and thus belongs to the group. It is also straightforward to check that the associativity condition is satisfied. There is an identity element (a rotation by zero degrees) and also an inverse for each element (the inverse of a rotation by angle θ is a rotation by angle $-\theta$). Thus, phase rotations form a group. Similar geometric considerations apply for the three-dimensional complex rotations associated with quark fields.

As well as the geometric arguments given earlier, we can also check the group properties algebraically, using the specific forms for how elements of each group act on the electron and quark fields. We saw in Eq. (2.36) that phase transformations act at each point in spacetime through multiplication by a unit complex number. The set of all unit complex numbers, combined according to standard multiplication, indeed satisfies the properties of a group. Combination of two elements gives

$$e^{i\alpha_1(x)}e^{i\alpha_2(x)} = e^{i(\alpha_1(x)+\alpha_2(x))},$$

which is also a unit complex number. Associativity amounts to the statement

$$e^{i(\alpha_1+\alpha_2)}e^{i\alpha_3} = e^{i\alpha_1}e^{i(\alpha_2+\alpha_3)},$$

which is indeed true. Finally, the identity element is 1, and the inverse of $e^{i\alpha(x)}$ is $e^{-i\alpha(x)}$. For colour rotations, let us recall that each $\mathbf{U}(x)$ is a 3×3 special unitary matrix. The product of any two such matrices is also a special unitary matrix. Furthermore, the combination rule for the transformations amounts to conventional matrix multiplication, which is known to be associative. The identity element is the 3×3 identity matrix $\mathbf{I}$, and the inverse of element $\mathbf{U}(x)$ is $\mathbf{U}^{-1}(x)$.

There are commonly used mathematical notations for the groups that we have encountered earlier. For example, SU(N) is defined to be the group corresponding to all special unitary matrices of dimension N under matrix multiplication. Thus, the colour rotations earlier form the group SU(3). We can use a similar notation for phase rotations: unit complex numbers count as 1×1 unitary (but not special unitary) matrices, which may seem like a bizarrely obtuse way of thinking about them. Nevertheless, it is common to see the earlier group of phase rotations labelled as U(1).

Note that both of the earlier groups have infinitely many elements, corresponding to the fact that there are continuous variables (e.g. rotation angles) that label each group element. Such groups are called *Lie groups* and are distinguished from *finite groups* of discrete transformations, such as reflections. A very large amount is known about Lie groups, and we can rely on this knowledge to arrive at an appropriate definition of the field strength for the gluon field. First, let us note that although the number of elements of a continuous (Lie) group may be infinite, we typically only need a finite number of parameters to specify any given element. In the case of phase rotations, a single parameter sufficed at each point in space-time, namely $\alpha(x)$. For the colour arrow rotations, this number will be the number of degrees of freedom in a general special unitary matrix of dimension 3. A general such matrix of dimension N has $N^2 - 1$ parameters.[6] Thus, a particular SU(3) transformation needs $3^2 - 1 = 8$ parameters to fully characterise it. To see how to express these parameters mathematically, a result commonly known as *Lie's theorem* allows us to write *any* element of a continuous group as

$$\mathbf{U} = \exp\left[i\sum_{a=1}^{D}\alpha^a\mathbf{T}^a\right], \tag{2.59}$$

where D is the dimension of the group (i.e. the number of parameters associated with each transformation). Here we have introduced D matrices $\{\mathbf{T}^a\}$, known as the *generators* of the group, and such that the exponential in Eq. (2.59) is defined by its Taylor expansion. Furthermore, the $\{\alpha^a\}$ are parameters accompanying each generator, and we have in fact already seen a special case of this result, namely the phase transformations of Eq. (2.36). There is only a single parameter in that case, which we labelled by α with no need for an index. Furthermore, the generator can be taken to be the number 1. Another example that may be familiar from your undergraduate days is that of rotations of spin states in quantum mechanics. You may recall that, for a spin-1/2 system, one may represent the effect of a rotation by the matrix

$$\mathbf{U} = \exp\left[i\sum_{a=1}^{3}\theta^a\tau^a\right], \quad \tau^a = \frac{\sigma^a}{2}, \tag{2.60}$$

where $\{\sigma^a\}$ are the *Pauli matrices* and $\{\theta^a\}$ the rotation angles around each spatial axis. This has precisely the form of Eq. (2.59), such that we may interpret the matrices $\{\tau^a\}$ as generators of spin-1/2 rotations. Incidentally, these form the group SU(2), given that they act on two-dimensional complex spin vectors. If you have seen this before, you may recall that the generators τ^a can be interpreted

---

[6]A general complex $N \times N$ matrix has $2N^2$ parameters, counting real and imaginary parts of each element. The unitary conditions constitute N constraints, and the determinant condition imposes a further single constraint.

as infinitesimal rotations about the appropriate axes. Likewise, from Taylor expanding Eq. (2.59) in the case where the parameters $\{\alpha^a\}$ are small, we can interpret $\mathbf{T}^a$ as infinitesimal transformations for any Lie group.

Explicit forms of the generators for a given Lie group can be chosen, although this choice is not unique in that any linear combination of the generators can be taken, which merely redefines the parameters $\{\alpha^a\}$. There is a consistency condition on the generators, though, which follows from the fact that the product of any two group elements must itself be an element of the group. A standard result known as the *Baker–Campbell–Hausdorff formula* tells us that a product of two exponentiated matrices yields

$$\exp\left[\mathbf{A}\right]\exp\left[\mathbf{B}\right] = \exp\left[\mathbf{A} + \mathbf{B} + \frac{1}{2}[\mathbf{A},\mathbf{B}] + \cdots\right], \qquad (2.61)$$

where we have made use of the *matrix commutator*

$$[\mathbf{A},\mathbf{B}] = \mathbf{A}\mathbf{B} - \mathbf{B}\mathbf{A}, \qquad (2.62)$$

and the ellipsis in Eq. (2.61) denotes terms involving multiple nested commutators of $\mathbf{A}$ and $\mathbf{B}$. Applying Eq. (2.61) to the product of two group elements of the form of Eq. (2.59), we will end up with a single exponential involving many different commutators of different generators. However, we must end up with another group element, which must necessarily have the *same* form as Eq. (2.59). The only way this can possibly be true is if the commutator of two generators is itself a superposition of generators:

$$[\mathbf{T}^a,\mathbf{T}^b] = \sum_c if^{abc}\mathbf{T}^c. \qquad (2.63)$$

Here the factor of i is conventional, and the parameters f^{abc} are known as *structure constants*. For a given Lie group and choice of generators, they are completely determined. Note that Eq. (2.63) shows that the structure constants must be antisymmetric under interchange of the indices a and b. It is actually possible to go further than this and show that they are completely antisymmetric:

$$f^{abc} = f^{cab} = f^{bca} = -f^{bac} = -f^{acb} = -f^{cba}. \qquad (2.64)$$

As an example, it is well known that the earlier matrices $\{\tau^a\}$ (related to the Pauli matrices) satisfy the relation

$$[\tau^a, \tau^b] = i \sum_c \epsilon^{abc} \tau^c, \tag{2.65}$$

where ϵ^{abc} is the three-dimensional *Levi-Civita symbol*, which is completely antisymmetric with $\epsilon^{123} = 1$. From Eq. (2.60), we thus find that the structure constants of SU(2) are

$$f^{abc}\Big|_{\text{SU(2)}} = \epsilon^{abc}. \tag{2.66}$$

In general, Eq. (2.63) is known as the *Lie algebra* associated with a given Lie group. Up to some caveats that are irrelevant for our purposes, a Lie group can be completely determined from its Lie algebra, as perhaps is clear from Eq. (2.59): as generators, we can choose any matrices obeying the Lie algebra, and we may then exponentiate them to generate all possible group elements.

Later on, we make use of a useful identity for the structure constants. First, note the *Jacobi identity* for matrices:

$$[\mathbf{A}, [\mathbf{B}, \mathbf{C}]] + [\mathbf{C}, [\mathbf{A}, \mathbf{B}]] + [\mathbf{B}, [\mathbf{C}, \mathbf{A}]] = 0, \tag{2.67}$$

which is easily verified by multiplying out the commutators and cancelling pairs of terms. Applying this to a trio of generators $\{\mathbf{T}^a, \mathbf{T}^b, \mathbf{T}^c\}$ and using Eqs. (2.63) and (2.64), one finds the following relation for the structure constants:

$$\sum_d \left(f^{bcd} f^{dea} + f^{abd} f^{dec} + f^{cad} f^{deb} \right) = 0. \tag{2.68}$$

Above, we have used a summation convention for spacetime indices, where explicit summation signs are omitted. From now on, we apply a similar convention to the indices associated with the Lie algebra, which we refer to as *colour indices* $\{a, b, c, \ldots\}$, to distinguish them from *spacetime indices* $\{\mu, \nu, \alpha, \ldots\}$. Finally, we note a further useful identity obeyed by the structure constants:

$$f^{abc} f^{a'bc} = T_A \delta^{aa'}, \tag{2.69}$$

which defines the normalisation constant T_A. It is straightforward to verify Eq. (2.69) for the group SU(2), and you should find $T_A = 2$.

2.3.2 Gluon field equation

In the previous section, we have seen that gauge transformations can be associated with Lie groups, which have certain mathematical properties. It follows that the global gauge transformation matrices $\mathbf{U}$ of Eq. (2.49) can be expressed in terms of generators as in Eq. (2.59), where the parameters $\{\alpha^a\}$ will be constant. For local gauge transformations, the generators will be the same at different points in space, as we are talking about the same type of transformation everywhere in spacetime. However, the parameters labelling each group element will become functions of position: $\alpha^a \to \alpha^a(x)$. Let us now use this knowledge to write down a field strength tensor for the gluon field $\mathbf{A}_\mu$ and ultimately its equation of motion.

First, let us return to the relative simplicity of the electromagnetic case and interpret the covariant derivative of Eq. (2.41) in more detail. For reasons that should hopefully become clear, let us consider the action of a certain operator on the electron field $\Psi(x)$:

$$e^{a^\mu D_\mu} \Psi(x) = e^{a^\mu \partial_\mu + iea^\mu A_\mu} \Psi(x). \qquad (2.70)$$

Here we have taken a constant 4-vector a^μ and contracted it with the covariant derivative, which we may interpret as "taking the covariant derivative in the direction a^μ". We have then exponentiated this operator and can interpret each term on the right-hand side as follows. The first term generates a translation according to the well-known result:

$$e^{a^\mu \partial_\mu} f(x) = f(x + a), \qquad (2.71)$$

as may be verified by Taylor expansion of both sides. The second contribution of the right-hand side of Eq. (2.70) corresponds to a gauge transformation, with parameter

$$\alpha(x) = ea^\mu A_\mu. \qquad (2.72)$$

Upon expanding the operator in Eq. (2.70) to linear order in a^μ, we get

$$e^{a^\mu D_\mu} = 1 + a^\mu D_\mu + \cdots = 1 + a^\mu(\partial_\mu + ieA_\mu) + \cdots, \qquad (2.73)$$

and it follows that we may interpret the first and second terms in the covariant derivative as generating an infinitesimal translation, plus an infinitesimal gauge transformation.

There is also a nice way to reinterpret the field strength tensor. Consider applying two covariant derivatives to Ψ and then taking the difference. This may be written as

$$
\begin{aligned}
[D_\mu, D_\nu]\Psi(x) &= [(\partial_\mu + ieA_\mu)(\partial_\nu + ieA_\nu) - (\mu \leftrightarrow \nu)]\Psi \\
&= [\partial_\mu\partial_\nu\Psi + ieA_\mu\partial_\nu\Psi + ie(\partial_\mu A_\nu)\Psi + ieA_\nu\partial_\mu\Psi \\
&\quad - e^2 A_\mu A_\nu\Psi - (\mu \leftrightarrow \nu)] \\
&= ie(\partial_\mu A_\nu - \partial_\nu A_\mu)\Psi(x).
\end{aligned}
\tag{2.74}
$$

Comparison with Eq. (2.46) allows us to recognise the field strength tensor in the final line, and if this equation is to be true for arbitrary electron fields $\Psi(x)$, then we must have

$$
F_{\mu\nu} = -\frac{i}{e}[D_\mu, D_\nu].
\tag{2.75}
$$

It is now straightforward to generalise the earlier remarks to the case of the gluon field. As in Eq. (2.41), the second term in Eq. (2.55) must generate infinitesimal gauge transformations. This in turn means that the field $\mathbf{A}_\mu$ must be a superposition of the generators $\{\mathbf{T}^a\}$ so that we can write

$$
\mathbf{A}_\mu(x) = A_\mu^a(x)\mathbf{T}^a,
\tag{2.76}
$$

a decomposition that is so widely used that you often see A_μ^a being referred to as "the gluon field". It is really 8 separate fields (for SU(3)) that are combined into a single object. Note that, in this language, the gluon field has one spacetime index and one colour index.

We may generalise the field strength tensor by simply forming the analogue of Eq. (2.75):

$$
\mathbf{F}_{\mu\nu} = -\frac{i}{g}[\mathbf{D}_\mu, \mathbf{D}_\nu],
\tag{2.77}
$$

such that an explicit calculation yields

$$
\mathbf{F}_{\mu\nu} = \partial_\mu\mathbf{A}_\nu - \partial_\nu\mathbf{A}_\mu + ig[\mathbf{A}_\mu, \mathbf{A}_\nu].
\tag{2.78}
$$

We can see immediately that this is more complicated than Eq. (2.46). In particular, whereas the electromagnetic field strength is linear in the field, the gluon field strength is non-linear. Another crucial difference – which is relevant for constructing an equation of motion for the field – is that the electromagnetic field strength tensor is locally gauge-invariant, whereas the gluon field strength is not. To see why, note that the transformation law of Eq. (2.54) implies

$$\mathbf{D}_\mu \Psi \to [\mathbf{U}\mathbf{D}_\mu \mathbf{U}^{-1}](\mathbf{U}\Psi).$$

The quantity in round brackets is the gauge-transformed field, which in turn implies that the covariant derivative itself transforms as

$$\mathbf{D}_\mu \to \mathbf{U}\mathbf{D}_\mu \mathbf{U}^{-1}. \tag{2.79}$$

From Eq. (2.77), we can then see that the gluon field strength transforms in the same way:

$$\mathbf{F}_{\mu\nu} \to \mathbf{U}\mathbf{F}_{\mu\nu}\mathbf{U}^{-1}. \tag{2.80}$$

The right-hand side is not the same as the left-hand side in general so that the field strength tensor is indeed not gauge-invariant. This is not a problem for setting up an equation of motion, which does not have to be composed of purely gauge-invariant objects. All that is required is that the equation of motion be gauge *covariant*, meaning that both sides transform nicely under gauge transformations. We see how this works shortly, but first note that Eq. (2.80) is a different sort of transformation to that of Eq. (2.52) for the quark field. Mathematically, one says that the field strength transforms in the *adjoint representation* of the gauge group and that the quark field transforms in the *fundamental representation*. This in turn means that we can write a covariant derivative for the field strength but that it will be different to its counterpart acting on quark fields. To this end, we may generalise the earlier remarks for electromagnetism, namely that the covariant derivative generates an infinitesimal translation plus an infinitesimal gauge transformation. A gauge transformation in the present case has the form

$$\mathbf{U} = \exp\left[ig\boldsymbol{\alpha}\right], \quad \boldsymbol{\alpha} = \sum_a \alpha^a \mathbf{T}^a$$

so that substituting this into Eq. (2.80) and expanding to linear order in g yield

$$\mathbf{F}_{\mu\nu} \to \mathbf{F}_{\mu\nu} + ig\,[\boldsymbol{\alpha}, \mathbf{F}_{\mu\nu}] + \mathcal{O}(g^2). \qquad (2.81)$$

The second term on the right-hand side represents the effect of an infinitesimal gauge transformation of the field strength. Following Eq. (2.72), we may write the gauge parameter as

$$\boldsymbol{\alpha} = a^\rho \mathbf{A}_\rho. \qquad (2.82)$$

Then, the effect of combining an infinitesimal translation with our gauge transformation takes the form

$$\left[a^\rho \tilde{\mathbf{D}}_\rho\right] \mathbf{F}_{\mu\nu}, \quad \tilde{\mathbf{D}}_\rho \mathbf{F}_{\mu\nu} = \partial_\rho \mathbf{F}_{\mu\nu} + ig\,[\mathbf{A}_\rho, \mathbf{F}_{\mu\nu}]. \qquad (2.83)$$

We have thus obtained the covariant derivative $\tilde{\mathbf{D}}_\mu$ that acts on the field strength,[7] or indeed on any quantity that transforms as in Eq. (2.80), given that this is the only property of $\mathbf{F}_{\mu\nu}$ that we have used in deriving Eq. (2.83). In Eq. (2.76), we introduced components $A_\mu^a(x)$ of the gauge field. We can do the same thing for the field strength:

$$\mathbf{F}_{\mu\nu} = F_{\mu\nu}^a \mathbf{T}^a, \qquad (2.84)$$

such that substituting Eqs. (2.76) and (2.84) into Eq. (2.83) and using Eq. (2.63), we obtain

$$\tilde{\mathbf{D}}_\rho \mathbf{F}_{\mu\nu} = \left[\partial_\rho F_{\mu\nu}^a - g f^{abc} A_\rho^b F_{\mu\nu}^c\right] \mathbf{T}^a, \qquad (2.85)$$

where the square brackets contain the components of the covariant derivative of the field strength.

We are now able to construct an equation of motion for the gluon field, returning to the earlier observation that we need an equation

---

[7]Many books use the same symbol for the covariant derivatives $\mathbf{D}_\mu$ and $\tilde{\mathbf{D}}_\mu$, despite the fact that these act on quantities in different representations of the gauge group. Here, we adopt the tilde notation to avoid confusion.

of motion involving the field strength, that transforms nicely under gauge transformations. It is straightforward to see that the quantity

$$\partial^\nu \mathbf{F}_{\mu\nu}$$

does not transform nicely, as upon using Eq. (2.80), one would generate derivatives of $\mathbf{U}$ or $\mathbf{U}^{-1}$ upon performing a gauge transformation. However, we can instead replace the partial derivative with our covariant derivative $\tilde{\mathbf{D}}_\mu$, obtaining the vacuum field equation

$$\tilde{\mathbf{D}}^\nu \mathbf{F}_{\mu\nu} = 0, \tag{2.86}$$

which will be preserved under gauge transformations by construction. To be fully general, we can include a source term on the right-hand side, as in the electromagnetic field equation of Eq. (2.30):

$$\tilde{\mathbf{D}}^\nu \mathbf{F}_{\mu\nu} = \mathbf{J}_\nu, \quad \mathbf{J}_\nu = J_\nu^a \mathbf{T}^a, \tag{2.87}$$

where the current is not itself gauge-invariant but must transform appropriately so as to match the left-hand side. Antisymmetry of the field strength, however, implies the *covariant conservation equation*

$$\tilde{\mathbf{D}}^\mu \mathbf{J}_\mu = 0, \tag{2.88}$$

which is a generalised form of the conservation equation of Eq. (2.29). Note also that Eq. (2.77) implies that the gluon field strength satisfies

$$\tilde{\mathbf{D}}_\mu \mathbf{F}_{\nu\alpha} + \tilde{\mathbf{D}}_\nu \mathbf{F}_{\alpha\mu} + \tilde{\mathbf{D}}_\alpha \mathbf{F}_{\mu\nu} = 0. \tag{2.89}$$

This is a generalisation of Eq. (2.31) and is also referred to as the *Bianchi identity*. Equations (2.87) and (2.89) completely determine how the gluon field is generated by colour charges and currents and are the analogue of the Maxwell equations for electromagnetism. One may also write a generalisation of the Lorentz force equation of Eq. (2.33), although we postpone a discussion of this until Chapter 7.

Equations (2.58), (2.87), and (2.89) form the basis of the theory of quarks and gluons, known as *Quantum Chromodynamics (QCD)*. All we have to do to complete the theory, in fact, is to note that there are six different types of quarks, such that we must include multiple fields obeying Eq. (2.58), where the mass is different in each case. However, if we want to, we can consider the theory without any quarks at all, such that we have a pure gluon field and nothing else. This is called *Yang–Mills theory* and was remarkably discovered decades before it was realised that it could describe the gluon [23]. The theory is completely defined by Eqs. (2.86) and (2.89) (or Eq. (2.86) if we want to include additional sources for the gluon field). We could also have done this for electromagnetism, where the equivalent is to consider only the photon field. However, the resulting theory is not particularly interesting: photons carry no electromagnetic charge, so that taking the electron field out leaves a theory of non-interacting photons, with no means of generating them. By contrast, the non-linearity of the Yang–Mills equations means that the gluon field interacts with itself. In other words, this means that gluons themselves carry colour charge, which is the physical meaning of the index a associated with the gluon field A_μ^a. Indeed, the interacting nature of the field is perhaps made clearer by explicitly computing the components of the field strength, as defined in Eq. (2.84). Substituting Eq. (2.76) into Eq. (2.78), one finds

$$F_{\mu\nu}^a = \partial_\mu A_\nu^a - \partial_\nu A_\mu^a - g f^{abc} A_\mu^b A_\nu^c. \tag{2.90}$$

We see that the non-linear term in the field strength tensor couples together two gluon fields with colour indices b and c, where the structure constants f^{abc} enter this interaction. Furthermore, substituting Eq. (2.76) into Eq. (2.87) ultimately leads to the following equation of motion for A_μ^a:

$$\partial^\nu F_{\mu\nu}^a - g f^{abc} A^{\nu b} F_{\mu\nu}^c = J_\nu^a. \tag{2.91}$$

This contains terms which are both quadratic and cubic in the gluon field, where again the structure constants are involved in coupling together fields with different colour indices. There are also explicit factors of g (or even g^2) in the gluon interaction terms, which allow us to interpret g as the strength of the interaction between gluons. From expanding the covariant derivative in Eq. (2.58), g is also found to be the strength of the interaction between quarks and gluons. It is called the *coupling constant* and is the analogue of e in the electromagnetic case. If you go carefully through the earlier formulae, you will notice that the fact that the strength of the gluon self-interactions is the *same* as that between quarks and gluons is no accident but a consequence of gauge invariance.

Above, we have explicitly focused on the case of SU(3) transformations, which act on a quark field with 3 colours. However, careful scrutiny of the arguments reveals that a similar theory would have been obtained if SU(3) were replaced with an arbitrary Lie group. In other words, Eqs. (2.87) and (2.89) apply for any gauge group, provided one uses the appropriate structure constants and generators. It is common to call such a theory a *Yang–Mills theory* in all cases[8] and to refer to the field $\mathbf{A}_\mu$ as the *gluon*, even though this is not literally the gluon of QCD. However, if the gauge group can be anything, why are the formulae for electromagnetism so different to those for Yang–Mills theory? After all, the former should emerge as a special case of the latter if we take the gauge group to be U(1). That this indeed works can be seen as follows. First, we may note from above that there is only one generator (the number 1) for U(1), which clearly commutes with itself under multiplication. Thus, there are no non-zero structure constants for the Lie group U(1), and there is only one possible value for the colour indices denoted earlier so that we may neglect them. Consequently, the field strength components of Eq. (2.90) reduce to

$$F^a_{\mu\nu} \to F_{\mu\nu} = \partial_\mu A_\nu - \partial_\nu A_\mu,$$

and the covariant derivative of Eq. (2.83) reduces to

$$\tilde{\mathbf{D}}_\rho \mathbf{F}_{\mu\nu} \to \partial_\rho F_{\mu\nu}.$$

Thus, Eq. (2.91) simplifies to Eq. (2.30), and Eq. (2.89) to Eq. (2.31) as required. Finally, we may show that the complicated gauge transformation of Eq. (2.57) reduces to that of Eqs. (2.42) and (2.44) for the group U(1). To see this, we may explicitly write $\mathbf{U} = e^{i\alpha}$ in Eq. (2.57), which yields

$$A'_\mu = A_\mu - \frac{1}{g}\partial_\mu \alpha.$$

This indeed matches Eqs. (2.42) and (2.44) upon relabelling $g \to e$.

---

[8]Incidentally, the original Yang–Mills theory of Ref. [23] had an SU(2) gauge group.

There may be other gauge groups for which the structure constants vanish i.e. such that the order in which we multiply arbitrary group elements is unimportant. Such groups are called *abelian*, to distinguish them from *non-abelian* groups such as SU(3). For this reason, electromagnetism and Yang–Mills theory are referred to as *abelian* and *non-abelian gauge theories*, respectively, and we use this terminology throughout this book.

2.4   General Relativity

The last theory we need to review in this chapter is general relativity which, to date, represents our most complete description of gravity. As for the theories considered in the previous sections, the summary given here is no substitute for a full introduction to the theory, and we refer the reader to Ref. [21] for a pedagogical exposition. The most basic idea of the theory is that matter (more specifically, any distribution of mass/energy) *curves* spacetime, and that it is this curvature that appears as the force of gravity. This is a striking departure from the nature of spacetime in non-gravitational physics. We may think of spacetime in the latter case as like the hard wooden stage of a theatre, in which the principal actors – the fundamental matter and force particles – prance about, without affecting the boards upon which they tread. In general relativity, it is as if the stage has been replaced with the surface of a trampoline, that warps and stretches in response to the actors' presence, and subsequent movements. We depict this idea schematically in Figure 2.3, which shows a very massive object sitting in spacetime, which is curved as a result. A freely falling smaller object moving

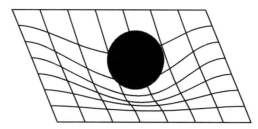

Fig. 2.3. A heavy object curves the spacetime around it, such that a smaller object will follow a curved path, that appears as a gravitational orbit.

in the spacetime will no longer follow a straight line (n.b. these no longer exist in a curved space) but will instead roll around the larger object, which we can recognise as a gravitational orbit. Note that Figure 2.3, whilst a useful visualisation, is a somewhat flawed analogy. In the full theory of GR, it is the full four-dimensional spacetime that is curved, which is not at all easy to visualise! In order to put this theory on a precise footing, we must do two things: (i) learn how to describe curved spacetimes mathematically and (ii) give a prescription for how the curvature of spacetime is related to the mass/energy distribution that is present. Let us take each of these in turn.

2.4.1 The mathematics of curvature

We have in fact already introduced some of the mathematical tools that we need, when we discussed the language of 4-vectors and tensors in *flat spacetime* (i.e. where gravity is absent), in Section 2.1. We saw in particular that one can measure spacetime distances near a given point using the quantity of Eq. (2.13), which is sometimes called the *line element*. This involved the metric tensor $\eta_{\mu\nu}$, whose components in Cartesian coordinates are the same at all points in spacetime, corresponding to the fact that spacetime is somehow "the same" everywhere. In a curved space, this is clearly no longer true, and the fact that spacetime may be more/less warped or stretched at different points means that the metric tensor will depend on position in general. Let us thus replace the flat-space metric with a more general metric tensor:

$$\eta_{\mu\nu} \to g_{\mu\nu}(x).$$

Note the fact that the metric depends upon spacetime position does not necessarily imply that spacetime is curved. As Eqs. (2.17) and (2.18) imply, transforming to a non-Cartesian coordinate system can make even the flat space metric depend upon position. We thus need a better way of characterising what it means to say that spacetime can be curved, and so it is useful to consider a three-dimensional, purely spatial, analogy. Consider the sphere shown in Figure 2.4, where we have constructed a triangle by taking two points A and B on the equator and joining them to each other and to a point C at the north pole. We have also shown a vector on

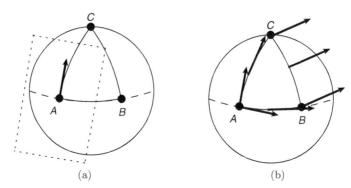

(a)   (b)

Fig. 2.4. (a) A vector on a sphere lives in a *tangent space*, namely a two-dimensional surface that is tangent to the sphere at that point; (b) parallel transport of a vector around a loop changes its orientation in general, in a curved space.

the sphere at point A, which is tangent to the curve AC. This vector lives in a *tangent plane*, namely a two-dimensional surface, all of whose directions are tangent to the sphere at the point A. A section of this plane is shown by the dotted lines in the figure, and one can clearly see that the tangent spaces in which vectors live are *different* at different points on the sphere. If you doubt this, buy yourself a suitably large orange (even better, a pomelo) and a piece of stiff paper and place the paper on the orange at different points on its surface, such that the paper is tangent to the orange. You will see you will have to change the orientation of the paper at different points on the orange, which amounts to constructing different tangent spaces.

In more pedestrian language, the earlier remarks imply that what we mean by a "vector" is different at different points in a curved space. This is very different to flat space, which can be characterised by the fact that the tangent spaces at different points are all similar. It means that we can easily compare vectors at different points in flat space, without having to think about it. In curved space, we have to think more carefully. If we want to take a vector at some point A and compare it with one at another point B, we need to somehow transport the vector from A to B so that it lies in the tangent space of B, rather than that of A. There are various ways of doing this, but one is the so-called *parallel transport*, in which we construct a curve from A to B and then define a rule for saying what it means for vectors to be kept parallel to themselves as they move along the curve. Consider, for example, transporting a vector V^μ from x^μ to $x^\mu + dx^\mu$, in a way that is not necessarily parallel. We may write the total change in the vector as

$$DV^\mu = [\partial_\nu V^\mu + \Gamma^\mu_{\alpha\nu} V^\alpha]\, dx^\nu, \qquad (2.92)$$

where the first term is the same as would be present in Cartesian coordinates in flat space. The second term thus represents the effect of the coordinate axes changing as we move from point to point, which may either be due to using curvilinear coordinates in flat space or the fact that the space is curved. This term must be linear in the infinitesimal displacement dx^μ, but it will also be linear in the vector V^μ: if we transport the sum of two vectors, this must be the same as transporting them individually and adding the transported vectors. The parameters $\Gamma^\mu_{\alpha\nu}$ are called *Christoffel symbols* and will depend on spacetime position in general. We see how to find them shortly, but first note that we can write Eq. (2.92) as

$$DV^\mu = (D_\nu V^\mu) dx^\nu, \quad D_\nu V^\mu = \partial_\nu V^\mu + \Gamma^\mu_{\alpha\nu} V^\alpha. \qquad (2.93)$$

The operator D_ν is called the *covariant derivative*, where the word "covariant" in this case means that it transforms properly as a tensor, according the transformation law of Eq. (2.23). Armed with the covariant derivative, we can *define* the parallel transport of a vector by the condition

$$D_\nu V^\mu = 0. \qquad (2.94)$$

To find the Christoffel symbols, imagine that we have a curve $x^\mu(\tau)$ parametrised by some parameter τ. Let us take two vectors $U^\mu(\tau)$ and $V^\mu(\tau)$ defined at the same point τ. We have yet to implement a condition that defines what it means for vectors to be parallelly transported, but one way of stating it is that upon transforming both vectors U^μ and V^μ in this manner, their dot product should not change. That is, we must have

$$\frac{d}{d\tau}\left[g_{\mu\nu}U^\mu V^\nu\right] = \left[(\partial_\alpha g_{\mu\nu})U^\mu V^\nu + g_{\mu\nu}(\partial_\alpha U^\mu)V^\nu\right.$$

$$\left. + g_{\mu\nu}U^\mu(\partial_\alpha V^\nu)\right]\frac{dx^\alpha}{d\tau}$$

$$= \left[\partial_\alpha g_{\mu\nu} - g_{\beta\nu}\Gamma^\beta_{\alpha\mu} - g_{\mu\beta}\Gamma^\beta_{\alpha\nu}\right]U^\mu V^\nu \frac{dx^\alpha}{d\tau} = 0,$$

$$(2.95)$$

where we have used the chain rule of partial differentiation and Eqs. (2.93) and (2.94). We have also relabelled contracted indices where necessary. It follows that the quantity in square brackets must vanish. This relates the Christoffel symbols to the metric tensor, and some rearrangement leads to the following expression[9]:

$$\Gamma^\beta_{\mu\nu} = \frac{1}{2}g^{\alpha\beta}\left[\partial_\mu g_{\nu\alpha} + \partial_\nu g_{\alpha\mu} - \partial_\alpha g_{\mu\nu}\right]. \qquad (2.96)$$

Having seen how to parallel transport vectors, let us now consider a specific example, which will in turn lead us to how to mathematically describe the curvature of spacetime. In Figure 2.4(b), we show a vector being transported in a parallel fashion around the triangular loop $A \to B \to C \to A$. Upon arriving back at the starting point, the vector has clearly been rotated by a right angle. This is a characteristic feature of curved spacetimes: parallel transport of vectors around closed loops will change their orientation in general. By taking such loops smaller and smaller, we can get a measure for the local curvature at a given point in spacetime. To see how this works, consider an infinitesimal loop γ around the point x^μ, such that the total change in a parallelly transported vector is

$$\Delta V^\mu = \oint_\gamma (\partial_\nu V^\mu)dx^\nu = -\oint_\gamma \Gamma^\mu_{\alpha\nu}V^\alpha dx^\nu, \qquad (2.97)$$

---

[9]In deriving Eq. (2.96), one must assume that the Christoffel symbols are symmetric in the lower indices ($\Gamma^\beta_{\mu\nu} = \Gamma^\beta_{\nu\mu}$). It is possible to consider also an antisymmetric part, referred to as the *torsion*, but this would result in an alternative theory to general relativity.

where we have used Eqs. (2.93) and (2.94) in the second equality. There are numerous ways to calculate this integral: one is to use the four-dimensional Stokes' theorem

$$\oint_\gamma \xi_\nu dx^\nu = \int\int_S (\partial_\rho \xi_\nu - \partial_\nu \xi_\rho) d\Sigma^{\nu\rho}, \tag{2.98}$$

where S is the area bounded by the loop, with area element[10] $d\Sigma^{\alpha\beta}$. One then finds

$$\Delta V^\mu = -\left[\partial_\rho(\Gamma^\mu_{\alpha\nu} V^\alpha) - \partial_\nu(\Gamma^\mu_{\alpha\rho} V^\alpha)\right] d\Sigma^{\nu\rho},$$

where we have removed the integral signs owing to the fact that the loop is infinitesimal. Expanding the derivatives and again using Eqs. (2.93) and (2.94), one finds

$$\Delta V^\mu = -R^\mu_{\sigma\rho\nu} V^\sigma d\Sigma^{\nu\rho}, \tag{2.99}$$

where

$$R^\mu_{\sigma\rho\nu} = \partial_\rho \Gamma^\mu_{\sigma\nu} - \partial_\nu \Gamma^\mu_{\sigma\rho} + \Gamma^\mu_{\alpha\rho}\Gamma^\alpha_{\nu\sigma} - \Gamma^\mu_{\alpha\nu}\Gamma^\alpha_{\rho\sigma}. \tag{2.100}$$

This is called the *Riemann curvature tensor* and would be zero in flat space, even if the Christoffel symbols are non-zero (e.g. due to having a non-Cartesian coordinate system). Note that this tensor does not capture all of the curved properties of a space: only those that can be determined locally. It is possible to have a space which is locally flat but *globally curved*. An example is the surface of a cylinder. One may easily convince oneself that parallel transport of a vector around any closed loop results in no rotation of the vector (Figure 2.5), and hence the Riemann curvature vanishes everywhere. However, the surface is clearly not the same as a flat plane!

---

[10]If we have a rectangular loop spanned by displacements dx_1^μ and dx_2^μ, then $d\Sigma^{\alpha\beta} = dx_1^\alpha dx_2^\beta - dx_2^\alpha dx_1^\beta$. This is a multi-dimensional analogue of the cross-product used to define an area element from two displacements $d\boldsymbol{x}_1$ and $d\boldsymbol{x}_2$ in 3-vector algebra.

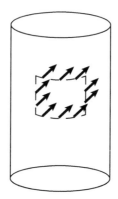

Fig. 2.5.   Parallel transport of a vector around a closed loop on the surface of a cylinder does not rotate the vector. The surface is *locally flat* but *globally curved*.

Interestingly, there is another definition of the Riemann tensor that one can give. One may verify from Eq. (2.93) that one has

$$[D_\rho, D_\nu]V^\mu \equiv (D_\rho D_\nu - D_\nu D_\rho)V^\mu = R^\mu{}_{\sigma\rho\nu}V^\sigma, \qquad (2.101)$$

such that the Riemann tensor can be defined in terms of the commutator of covariant derivatives. Furthermore, one may show that it obeys the *Bianchi identity*

$$D_\alpha R_{\mu\nu\beta\gamma} + D_\gamma R_{\mu\nu\alpha\beta} + D_\beta R_{\mu\nu\gamma\alpha} = 0. \qquad (2.102)$$

The similarity with the field strength tensor and Bianchi identity in Yang–Mills theory may not have gone unnoticed by the reader. Indeed, the geometric interpretation of these structures in the Yang–Mills case can be put on a much firmer footing: see e.g. Ref. [24] for a pedagogical exposition.

From the Riemann tensor, one may form quantities with lower numbers of spacetime indices. First is the *Ricci tensor*, defined by the contraction

$$R_{\sigma\nu} = R^\mu{}_{\sigma\mu\nu}, \qquad (2.103)$$

which turns out to be symmetric ($R_{\sigma\nu} = R_{\nu\sigma}$). Finally, one may contract the remaining indices to make the *Ricci scalar*

$$R = R^\sigma{}_\sigma. \qquad (2.104)$$

2.4.2   The Einstein equations

Having seen how to describe the (local) curvature of spacetime, let us now see how general relativity relates this to the mass and energy distribution. First, let us note that the latter may be described by a quantity known as the *energy–momentum tensor* $T^{\alpha\beta}(x)$, which may be different at different spacetime points, and is a four-dimensional analogue of the stress tensor in continuum mechanics. The component $T^{\mu\nu}$ represents the flux of 4-momentum component μ in the ν-direction, from which it follows that the T^{00} and T^{0i} components represent the local energy and 3-momentum density, respectively, where Latin indices range from 1 to 3. The diagonal components T^{ii} represent pressures in each of the spatial i directions, and finally the various off-diagonal space-like components T^{ij} ($i \neq j$) correspond to shear stresses. Various results are known for the energy–momentum tensor of different types of stuff (e.g. astrophysical dust, fluids, point particles, and radiation). Furthermore, the energy–momentum tensor is symmetric ($T^{\mu\nu} = T^{\nu\mu}$), which follows from its definition.

The energy–momentum tensor must somehow be related to symmetric two-index tensors that are related to the curvature. There are two possibilities: (i) the Ricci tensor $R_{\mu\nu}$ and (ii) the Ricci scalar R multiplied by the (symmetric) metric tensor $g_{\mu\nu}$. It turns out that the particular relation that is required is the *Einstein field equation*

$$R_{\mu\nu} - \frac{1}{2}Rg_{\mu\nu} = 8\pi G_N T_{\mu\nu}, \qquad (2.105)$$

where G_N is Newton's constant. This was originally derived by appealing to the limit of weak gravitational fields, where Newton's theory of gravity must ultimately emerge. You may be worried that this does not uniquely fix the left-hand side of Eq. (2.105), a quantity known as the *Einstein tensor* $G_{\mu\nu}$. However, a result known as *Lovelock's theorem* [25] tells us that the left-hand side is the only possibility if one wants a local equation of motion containing no more than second derivatives of the metric tensor, up to the addition of an extra term:

$$G_{\mu\nu} + \Lambda g_{\mu\nu} = 8\pi G_N T_{\mu\nu}. \qquad (2.106)$$

The extra term involves a constant multiple of the metric, where Λ is known as the *cosmological constant*. The physical interpretation is that Λ represents a constant energy density filling all space, and experiments tell us that it is indeed non-zero but very small. Often, however, we neglect the cosmological constant and consider the pure Einstein equation of Eq. (2.105).

Equation (2.105) is the equivalent of Maxwell's equations that tell us how the electromagnetic field A_μ is generated by a current density j^μ. We also need the analogue of the Lorentz force law of Eq. (2.33) that in the present case must tell us how test particles behave in a gravitational field. To this end, we may rely on the crucial insight of GR, namely that the curvature of spacetime is such as to create the gravitational force. It follows that particles experiencing no other forces follow *geodesics* in spacetime, namely those curves that are the equivalent of straight lines in flat spacetime. One definition that one can give of such curves is that they represent curves of extremal distance between any two points. There is another definition that can be shown to be equivalent, however, that is convenient for our purposes: a geodesic is such that the tangent vector to the curve is parallelly transported along it. This is clearly true for straight lines in flat space. Furthermore, we can see that it is also true for the vector that is transported along the line AC in Figure 2.4(b), where this line happens to be a geodesic. Given a curve $x^\mu(\tau)$, its tangent vector will be $dx^\mu/d\tau$, such that the parallel transport condition becomes

$$\frac{d^2 x^\mu}{d\tau^2} + \Gamma^\mu_{\alpha\beta} \frac{dx^\alpha}{d\tau} \frac{dx^\beta}{d\tau} = 0. \tag{2.107}$$

This is called the *geodesic equation* and will apply to the trajectory $x^\mu(\tau)$ of a freely falling particle, with τ its proper time.

In this chapter, we have reviewed some of the main theories of fundamental physics that we will study in more detail throughout the rest of this book. Although there are some similarities, there seems to be a key distinction between (non-)abelian gauge theories on the one hand and gravity on the other. In particular, gauge theories describe particles living in (flat) spacetime, whereas gravity constitutes *spacetime itself*, as represented by the metric or curvature tensors. On such grounds, it is not at all clear that the theories should be related to each other. Remarkably, however, they are, and we start to explore how this works in the following chapter.

Chapter 3

The Double Copy for Scattering Amplitudes

In this chapter, we start to explore the intriguing relationships between gauge, gravity, and other theories that form the main subject of this book. Chief amongst these is the *double copy* which, as mentioned in Chapter 1, began life as a relationship between *scattering amplitudes* in gauge and gravity theories [2–4]. This was itself inspired by previous work in string theory [1], and we examine the main ideas of this connection in what follows. First, we clearly have to review what a scattering amplitude is, and why we might care about this in the first place.

3.1 Scattering Amplitudes

Our main way of testing theories of fundamental physics is to use particle accelerators that collide focused beams of particles, before collecting the resulting debris in large detectors for subsequent analysis. Each collision constitutes a *scattering event* that takes an initial state of two beam particles and produces a final state. The latter may contain the incoming particles (possibly with different momenta), or it may contain completely different particles, due to creation or destruction of particles during the intermediate interaction. All of the theories underlying current accelerator experiments are necessarily quantum and relativistic, and thus the machinery of quantum field theory is needed to gain a full understanding of scattering processes. However, the concept of scattering occurs just as widely in non-relativistic quantum mechanics as well as in classical physics. In gravity, for example, the types of events that cause detectable gravitational wave emission are essentially scattering processes involving highly massive objects, such as black holes or neutron stars. Even in these classical examples, the ideas of quantum field theory can be useful, as we see later on.

Although a full treatment of quantum field theory would fill many more pages than this book can provide, the basic ideas of how scattering is described are not too complicated to state. First of all, let us note that in a quantum theory, it is not possible to say exactly what will happen in any given scattering event, but instead we can only calculate a probability. A generic scattering event will look something like Figure 3.1, where two beam particles come in with 4-momenta $\{p_1, p_2\}$ and a number of other particles with momenta $\{p_3, \ldots p_n\}$ emerge. Some sort of interaction happens in between, and it is typically assumed that this is localised in time so that the approximation of free, non-interacting particles at times $t \to \pm\infty$ is reasonable. In a quantum theory, we can label all possible incoming states by state vectors $\{|n; \text{in}\rangle\}$, where n denotes the number of particles, although we would also have to label their momenta, charges, polarisations, and any other quantum numbers to fully characterise each state. These states live in an abstract mathematical space called a *Fock space*, that is, the analogue of the *Hilbert space* that crops up in non-relativistic quantum mechanics. If you have

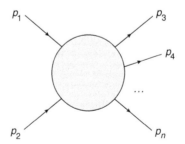

Fig. 3.1. In a scattering process, two incoming particles interact, producing a number of (possibly different) particles in the final state. Labels denote 4-momenta.

not seen these terms before, they do not matter: the only important thing is that we have a complete set of all possible initial states, labelled by all possible particle contents and properties. Likewise, all possible final states will correspond to some set $\{|m; \text{out}\rangle\}$. If we assume that the set of all possible outgoing particles is the same as the set of all possible incoming ones, it must then be true that the two sets of states $\{|n; \text{in}\rangle\}$ and $\{|m; \text{out}\rangle\}$ form two different bases for the *same* quantum state space. According to the rules of quantum mechanics, they must then be related by some unitary operator $\hat{S}$:

$$\hat{S}|n; \text{in}\rangle = |n; \text{out}\rangle, \quad \hat{S}^\dagger \hat{S} = \hat{S}\hat{S}^\dagger = \hat{I}, \tag{3.1}$$

where $\hat{I}$ denotes the identity operator. If we have a given initial state $|n; \text{in}\rangle$ and final state $|m; \text{out}\rangle$, another rule of quantum mechanics tells us that the probability of going from the first state to the second is given by

$$P_{nm} \propto |\langle m; \text{out}|n; \text{in}\rangle|^2, \tag{3.2}$$

which involves the overlap or *inner product* between the two states, whose modulus is then squared. Substituting Eq. (3.1), we find that scattering probabilities are given by the so-called *elements of the S-matrix*

$$S_{nm} = \langle m; \text{out}|\hat{S}^\dagger|n; \text{out}\rangle = \langle m; \text{in}|\hat{S}|n; \text{in}\rangle. \tag{3.3}$$

For this reason, $\hat{S}$ is called the *scattering operator*, and Eq. (3.3) shows that we may evaluate scattering probabilities in principle purely by considering states of free particles at $t = -\infty$ or $t = \infty$, provided we find a way to calculate the S-matrix elements.

It is conventional to decompose the scattering operator as follows:

$$\hat{S} = \hat{I} + i\hat{T}, \tag{3.4}$$

where $\hat{I}$ represents the identity operator in the space of quantum states, corresponding to the final state being exactly the same as the initial state, and thus nothing happening. The second term contains the *transition operator* $\hat{T}$, which instead describes something interesting happening. Switching to the more pedestrian notation of $|i\rangle$

and $|f\rangle$ for our initial and final states in some basis (in or out), we may then write elements of the S-matrix as

$$S_{fi} = \delta_{fi} + iT_{fi}, \tag{3.5}$$

where the transition matrix elements in the second term are further decomposed as

$$T_{fi} = (2\pi)^4 \delta^{(4)}(P_f - P_i)\mathcal{A}_{fi}. \tag{3.6}$$

Here the numerical factors are conventional, and the delta function imposes overall 4-momentum conservation between the initial and final states (with respective momenta P_i and P_f), which must necessarily be present on general grounds. The remaining quantity $\mathcal{A}_{fi}$ is where all the interesting stuff sits! It is called the *scattering amplitude* and will depend on the initial and final states in general. For given states, it is a single complex function of the momenta, polarisations, charges, and other quantum numbers, and it is the job of QFT (or some alternative theory) to tell us how to calculate it.

3.2 Amplitudes in Yang–Mills Theory

The S-matrix elements that we saw in the previous section are related to probabilities for particles to interact and thus to all of the physically observable quantities that particle physicists measure (such as cross-sections, decay rates, and distributions of particle properties). As Eq. (3.5) makes clear, the problem of determining S-matrix elements is entirely equivalent to being able to work out the scattering amplitude for a given initial and final state. Whilst the rules of QFT tell us in principle how to do this, it is typically not true that we can calculate any amplitude exactly. Rather, we must rely on approximate calculations, which can be justified if the strength of the interaction we are talking about is sufficiently weak. More precisely, we saw that all of the theories in Chapter 2 involved some number called the *coupling constant* that determines the strength of the force. For electromagnetism and Yang–Mills theory, these were labelled as e and g, respectively. For gravity, the equivalent number is related to Newton's constant G_N, as we see later on. Approximate calculations of scattering amplitudes mean that we calculate these as an expansion in the coupling, assuming this to be sufficiently small.[1] This is called *perturbation theory*, and the coefficients of the perturbation expansion will be complicated functions of particle properties (e.g. momenta, polarisations, and charges).

QFT has been with us for almost a century at the time of writing, and we thus have many methods for calculating scattering amplitudes in perturbation theory. Recent years, however, have seen an astonishing growth in new approaches, using complex mathematical and physical ideas to make such calculations as efficient as possible: see Refs. [26–28] for comprehensive reviews. For our purposes here, however, we review more traditional methods, in order to make contact with concepts with which the reader is more likely to be familiar. Indeed, in Chapter 1, we showed how *Feynman diagrams* can be handy pictorial representations of scattering processes in space-time. They are in fact much more than this: each diagram can be translated into a precise mathematical expression using the so-called *Feynman rules*, associated with each vertex, internal line, or external line. At a given order in perturbation theory, one must draw all possible Feynman diagrams with the desired initial and final states, translate them into algebra using the Feynman rules, and then sum the contributions together to get the total scattering amplitude.

A full derivation of the Feynman rules for a given theory is clearly beyond the scope of this book, and we refer the reader to a suitable QFT textbook, examples of which include Refs. [29–35]. However, provided the reader is prepared to take them on trust, it is relatively simple to state the Feynman rules for Yang–Mills theory, with which we are concerned in this section. There are a couple of things to note before we proceed. First, one may quote the Feynman rules in either *momentum space* or *position space*, where the results are mutually related by (inverse) Fourier transformation in four spacetime dimensions. We work directly in momentum space, given that scattering experiments typically have incoming beams with fixed 4-momenta. Second, Yang–Mills theory is a non-abelian gauge theory, meaning

---

[1]In fact, the coupling of constant of QCD is not small at low energies, which is related to how quarks and gluons are confined within hadrons. However, quantum corrections make the effective value of the coupling constant change with energy scale so that the QCD coupling constant can indeed be taken to be small when calculating relevant quantities for contemporary collider experiments.

that there are infinitely many ways in which one can represent the gluon field for a given physical solution. This makes the Feynman rules ambiguous, and one may fully specify them by *fixing a gauge*, which means putting extra constraints on the gauge field. The mathematical results obtained from individual Feynman diagrams then depend upon the choice of gauge, but the total result for the scattering amplitude – which is directly related to physically measurable quantities – does not. Here, we adopt the *Feynman gauge*, whose technical definition can be found in the textbooks cited earlier, and which does not bother us here. Then the Feynman rules for calculating amplitudes in momentum space are as follows:

(1) For a given initial and final state, draw all possible connected Feynman diagrams involving the vertices shown in Figure 3.2. Note that gluons are conventionally depicted as curly lines and that each gluon line entering the vertex has a spacetime index μ associated with it, and a colour index a, corresponding to the fact that the corresponding gauge field (A_μ^a) also has these.

(2) In each Feynman diagram, 4-momentum must be conserved at every vertex. If there are internal lines whose momentum is not completely fixed by the external momenta, one must introduce additional momenta so that everything can be labelled. An example is shown in Figure 3.3, and we can convince ourselves that the additional momentum k arises because of the fact that there is a loop in the diagram. Thus, these additional momenta are called *loop momenta*, and there will be L such momenta if there are L distinct loops.

(3) After labelling all momenta in a given diagram (after choosing where to put the loop momenta if needed), one must associate with each internal line a factor

$$\Delta_{\mu\nu}(p^2) = -\frac{i\delta^{ab}\eta_{\mu\nu}}{p^2}, \qquad (3.7)$$

where p is the 4-momentum of the line, and the spacetime and momentum indices are labelled as in Figure 3.4. This is called the *propagator* and is different in different gauges.

The Double Copy for Scattering Amplitudes 59

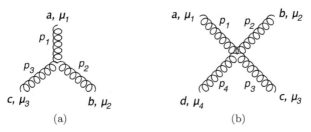

Fig. 3.2. (a) The 3-gluon vertex of Yang–Mills theory, corresponding to Eq. (3.8);
(b) the 4-gluon vertex, corresponding to Eq. (3.9). In both cases, all 4-momenta
are taken to be outgoing.

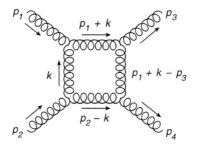

Fig. 3.3.   When Feynman diagrams involve loops, one must introduce extra *loop
momenta* such as k in order to be able to label the 4-momenta of all external and
internal lines. There is one loop momentum for every distinct loop, and we may
choose where to put it.

Fig. 3.4.   An internal gluon line in a diagram will have spacetime and colour
indices at both ends. In the Feynman gauge, each such line is associated with the
propagator of Eq. (3.7).

(4) Each 3-vertex (with momenta labelled as in Figure 3.2(a)) is
associated with a factor

$$V_3^{abc;\mu_1\mu_2\mu_3}(p_1, p_2, p_3) = -gf^{abc}\left[(p_1 - p_2)^{\mu_3}\eta^{\mu_1\mu_2}\right.$$
$$\left. + (p_2 - p_3)^{\mu_1}\eta^{\mu_2\mu_3} + (p_3 - p_1)^{\mu_2}\eta^{\mu_1\mu_3}\right].$$
$$(3.8)$$

Note that all momenta are outgoing in Figure 3.8. If they
are incoming instead, one should reverse the sign of the
4-momentum.

(5) Each 4-vertex (with momenta labelled as in Figure 3.2(b)) is associated with a factor

$$V_4^{abcd;\mu_1\mu_2\mu_3\mu_4}(p_1,p_2,p_3,p_4)$$

$$= -ig^2 \left\{ f^{eac}f^{ebd} \left[\eta^{\mu_1\mu_2}\eta^{\mu_3\mu_4} - \eta^{\mu_1\mu_4}\eta^{\mu_2\mu_3} \right] \right.$$

$$+ f^{ead}f^{ebc} \left[\eta^{\mu_1\mu_2}\eta^{\mu_3\mu_4} - \eta^{\mu_1\mu_3}\eta^{\mu_2\mu_4} \right]$$

$$\left. + f^{eab}f^{ecd} \left[\eta^{\mu_1\mu_3}\eta^{\mu_2\mu_4} - \eta^{\mu_1\mu_4}\eta^{\mu_2\mu_3} \right] \right\}. \qquad (3.9)$$

(6) Each external incoming gluon of 4-momentum p^μ is associated with a *polarisation vector* $\epsilon_\mu(p)$. These are any vectors that are transverse to p^μ ($\epsilon \cdot p = 0$) and a further reference vector, thus representing the two polarisation states of the gluon. A common choice is to choose the following basis of polarisation vectors, for left-circularly and right-circularly polarised gluons, respectively:

$$\epsilon_\mu^{\pm} = (0,1,\pm i,0). \qquad (3.10)$$

Each external outgoing gluon is associated with a conjugate polarisation vector $\epsilon_\mu^\dagger(p)$: reversing the 4-momentum interchanges the two circular polarisations, which indeed sends the vectors of Eq. (3.10) to their conjugates.

(7) Quantum mechanics tells us that we must always sum over equivalent ways in which a given final state can happen. Thus, we must sum (integrate) over all possible values of the loop 4-momenta. For a loop momentum k, this integral takes the form

$$\int \frac{d^4k}{(2\pi)^4},$$

where the denominator factors correspond to a convention in defining the Fourier transform to momentum space.

(8) Finally, one must divide by a *symmetry factor* for each Feynman diagram, corresponding to the dimension of the group of transformations that sends the diagram to itself. These symmetry factors are the subject of many arguments amongst theoretical physicists over conference lunches, and we need not worry about them too much in this book.

These are a lot of rules, but then QFT is a complicated theory! Despite first appearances – if this is indeed the first time you are seeing this stuff – Feynman rules revolutionised calculations in QFT, providing a systematic prescription for calculating the results of *any* scattering process, limited only by the computational power of humans or (super-)computers. We shortly see an example of applying the earlier rules, which may go a long way in helping to clarify them. Before this, let us make some preliminary remarks. First, we can note that rule 7 is familiar from Chapter 1, in which we talked about the need to sum over loop momenta when talking about how gravity is non-renormalisable. We now see how to formalise this in terms of an integral over loop momenta. The UV divergences of gauge theories and gravity emerge due to the fact that the integral does not converge in the region for which the components of the loop momentum are large. Next, let us note that the vertex factors contain explicit factors of the Yang–Mills coupling constant g. This means that, if we wish to work to a given order in perturbation theory only, we only have to include diagrams with a fixed number of vertices. Clearly, this dramatically simplifies the calculation, and it also explains why perturbation theory becomes increasingly difficult as the order in the coupling expansion increases: this corresponds to adding more external legs and/or loops, making the integrals needed to obtain the amplitude vastly more complicated.

Another interesting feature of the above Feynman rules is the propagator factor of Eq. (3.7). It contains an inverse power of the squared 4-momentum of the internal line, which when expanded in components $p^\mu = (E, \boldsymbol{p})$ is

$$\sim \frac{1}{E^2 - \boldsymbol{p}^2}.$$

Gluons are massless particles, and thus by the energy–momentum relation of Eq. (2.6), we would naïvely expect the denominator of the propagator to vanish. It does not, however, given that the 4-momentum of internal lines – as determined from momentum conservation involving external lines – has a non-zero square in general. We see this explicitly in the following example, but it is not a problem in practice given that the internal lines represent gluons that are never physically observed. We call them *virtual gluons* to distinguish them from the *real gluons* that go out to infinity, which is what the external lines represent. Another language that is commonly used is to describe real gluons as being *on the mass-shell*, or simply *on-shell* for short. Likewise, virtual gluons are *off-shell*, and the square of their 4-momentum is colloquially referred to as their *off-shellness*, or *virtuality*. When we integrate over loop momenta, there may be regions of the integration in which we are indeed sensitive to the vanishing of propagator denominators. To get round this, we may modify the expression of Eq. (3.7) to read

$$\Delta_{\mu\nu}(p^2) = -\frac{i\eta_{\mu\nu}}{p^2 - i\varepsilon}, \qquad (3.11)$$

where ε is an infinitesimal parameter. This is called the *Feynman prescription* and is such that we can interpret the loop integrals as contour integrals in the complex plane of the relevant momentum components. The sign of the parameter ε is not arbitrary and indeed can be shown to be crucial for making sure that QFT is *causal* (i.e. such that things that happen now can only affect things in the future). A full discussion can be found in the QFT books cited earlier [29–35].

Before discussing an example, let us also note that the earlier Yang–Mills Feynman rules, complicated enough as they are, are not quite complete! It turns out that in some gauge choices – including the Feynman gauge – the gluon field contains spurious degrees of freedom that need to be removed. Physically, we know that the gluon can only have two transverse polarisations which, for real particles, can be described by the polarisation vectors of Eq. (3.10). For virtual particles however, explicit calculation shows that a *longitudinal* polarisation mode contributes, which shouldn't be there. There is an ingenious fix for this problem, which amounts to adding so-called *ghost particles* into the theory, which couple to the gluon according to the vertex shown in Figure 3.5, and which carry colour charge in the same way. The corresponding Feynman rule is

$$V^{abc;\mu} = -gf^{abc}p^{\mu}, \qquad (3.12)$$

and we also need a propagator for an internal ghost line with colour indices a and b at its endpoints:

$$\Delta^{ab} = \frac{i\delta^{ab}}{p^2 - i\varepsilon}. \qquad (3.13)$$

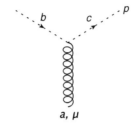

Fig. 3.5. Feynman rules for a ghost particle (dotted line) interacting with a gluon. Here (a, b, c) label colour indices and μ is a spacetime index. We also label the 4-momentum p of the ghost on the right-hand side.

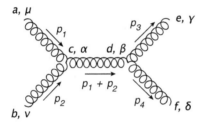

Fig. 3.6. A Feynman diagram contributing to $2 \to 2$ gluon scattering.

Finally, there is an additional stipulation that one should add a minus sign by hand if there is a loop of ghost particles in a given Feynman diagram. This minus sign ensures that the ghost diagrams precisely cancel the unwanted contributions, to all orders in perturbation theory. If this looks like a fudge – and it certainly will do to the uninitiated – then rest assured that there is a rigorous way of deriving the need for ghost particles [36], in arbitrary gauges.[2] We see these ghosts appearing in Chapter 5, and we see a different type of ghost in Chapter 7, which is interesting to compare with the earlier discussion. Furthermore, we have now given the interested reader all they need, at least in principle, to calculate Yang–Mills scattering amplitudes to arbitrary order in perturbation theory! Let us now turn to an example, namely the scattering of a pair of gluons to give another pair of gluons. We consider the diagram shown in Figure 3.6, which is indeed allowed, and is known as an *s-channel graph* in the literature on high-energy scattering. Applying the Feynman

---

[2]There are in fact gauge choices where the ghost particles are not needed. However, the price one pays is that the Feynman rules are much more complicated!

rules gives

$$
\mathcal{A}_s = \overbrace{\epsilon_\mu(p_1)\epsilon_\nu(p_2)}^{\text{Initial gluons}} \overbrace{\epsilon^\dagger_\gamma(p_3)\epsilon^\dagger_\delta(p_4)}^{\text{Final gluons}} \overbrace{\left(\frac{-i\eta_{\alpha\beta}\delta^{cd}}{(p_1+p_2)^2} \right)}^{\text{Internal line}}
$$

$$
\times \ (-gf^{acb}) \left[(-2p_1+p_2)^\nu \eta^{\mu\alpha} + (p_1+2p_2)^\mu \eta^{\alpha\nu} \right.
$$

$$
\left. - (p_2-p_1)^\alpha \eta^{\mu\nu} \right] \Big\} \leftarrow \text{First vertex}
$$

$$
\times \ (-gf^{def}) \left[(p_3-p_4)^\beta \eta^{\gamma\delta} - (2p_3-p_4)^\delta \eta^{\beta\gamma} \right.
$$

$$
\left. + (2p_4-p_3)^\gamma \eta^{\beta\delta} \right] \Big\} \leftarrow \text{Second vertex.} \tag{3.14}
$$

Here we have been careful to reverse the signs of the momenta if they are incoming to a vertex, and we have also used the overall momentum conservation condition

$$
p_1 + p_2 = p_3 + p_4 \tag{3.15}
$$

to write the 4-momentum of the internal line as p_3+p_4 where appropriate, leading to a slightly reduced expression. Equation (3.14) is a tangled mess of indices, structure constants, and tensors – I was not lying when I said that Yang–Mills is a complicated theory! However, it is possible to simplify things a little. First of all, we know that the polarisation vectors must satisfy

$$
\epsilon_i \cdot p_i = \epsilon^\dagger_i \cdot p_i = 0,
$$

using the simplified notation $\epsilon_i \equiv \epsilon(p_i)$. This allows us to get rid of many of the terms in Eq. (3.14). In order to write down what is left, we may introduce the so-called *Mandelstam invariants*

$$
s = (p_1+p_2)^2 = (p_3+p_4)^2,
$$

$$
t = (p_1-p_3)^2 = (p_2-p_4)^2,
$$

$$
u = (p_1-p_4)^2 = (p_2-p_3)^2. \tag{3.16}
$$

One then finds

$$\mathcal{A}_s = -\frac{ig^2}{s} f^{acb} f^{cef} \left[-2p_1 \cdot \epsilon_2 \, \epsilon_1^\alpha + 2p_2 \cdot \epsilon_1 \, \epsilon_2^\alpha + (p_1 - p_2)^\alpha \epsilon_1 \cdot \epsilon_2 \right]$$

$$\times \left[-2p_3 \cdot \epsilon_4 \, \epsilon_{3\alpha}^\dagger + 2p_4 \cdot \epsilon_3 \, \epsilon_{4\alpha}^\dagger + (p_3 - p_4)_\alpha \epsilon_3 \cdot \epsilon_4 \right]$$

$$= g^2 \frac{n_s c_s}{s}, \tag{3.17}$$

where in the second line we have isolated the overall coupling factor, the denominator, and the *colour factor*

$$c_s = f^{bac} f^{cef} \tag{3.18}$$

(n.b. we have shuffled indices about using Eq. (2.64) for later convenience). The *kinematic numerator* n_s collects the remaining mess of 4-momenta and polarisation vectors. Of course, this is only one possible diagram for the $2 \to 2$ scattering of gluons. The other diagrams are shown in Figure 3.7. The first two are conventionally called t-channel and u-channel diagrams, respectively, and each involves an internal line, whose squared 4-momentum yields one of the Mandelstam invariants in Eq. (3.16). They thus contribute to the amplitude in a similar fashion to Eq. (3.17):

$$\mathcal{A}_t = g^2 \frac{c_t n_t}{t}, \quad \mathcal{A}_u = g^2 \frac{c_u n_u}{u}, \tag{3.19}$$

where the relevant kinematic numerators can be obtained by relabelling momenta and polarisation vectors appropriately (see e.g. Ref. [29] for an explicit discussion of this point). The colour factors for the t- and u-channel diagrams are given by

$$c_t = f^{aec} f^{cfb}, \quad c_u = f^{fac} f^{ceb}, \tag{3.20}$$

which are easily obtained from Figures 3.7(a) and 3.7(b) recognising the clockwise pattern of colour indices in the 3-gluon vertex of Eq. (3.8). Note that we associated a single colour index with each internal line, rather than considering distinct colour indices for the endpoints: our experience with the s-channel diagram of Figure 3.6

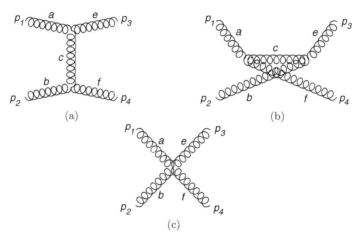

Fig. 3.7. (a) A t-channel diagram for gluon-gluon scattering; (b) a u-channel diagram; (c) 4-gluon vertex diagram.

taught us that the propagator merely equates the indices at both ends. Interestingly, the colour factors of Eqs. (3.18) and (3.20) are not independent. Comparison with the Jacobi identity of Eq. (2.68), and use of Eq. (2.64) where necessary, implies that

$$c_s = c_t + c_u, \qquad (3.21)$$

a fact that will turn out to be very significant in the following section.

The final contribution to the $2 \to 2$ scattering amplitude is the 4-gluon vertex graph of Figure 3.7(c), for which we must use the rule of Eq. (3.9). Rather than calculating this explicitly, let us merely note that – after relabelling colour indices as appropriate – the three separate terms in Eq. (3.9) correspond to the colour factors we have already seen in Eqs. (3.18) and (3.20)! Thus, the effect of the additional diagram is merely to shift the kinematic numerators we have already discussed. It follows that the final result for the total amplitude has the following form:

$$\mathcal{A} = g^2 \left(\frac{c_s n_s}{s} + \frac{c_t n_t}{t} + \frac{c_u n_u}{u} \right), \qquad (3.22)$$

where we have absorbed the 4-gluon vertex contributions into the existing numerators, rather than introduce a new notation. There is a nice way to think about this expression. The structure of the denominators suggests we can consider only the *effective diagrams*

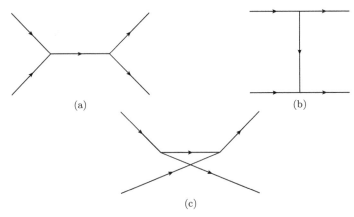

(b)

(c)

Fig. 3.8.   Effective diagrams associated with the expression of Eq. (3.22), where diagrams (a), (b) and (c) corresponds to the first, second and third terms, respectively.

with 3-vertices shown in Figure 3.8 but where the "rules" for obtaining the kinematic numerators are not the usual Feynman rules given earlier. We draw these effective diagrams using solid lines for the gluons rather than the usual curly symbols to make clear that these are not Feynman diagrams in the strict sense. Rather, all 4-vertices have been gotten rid of by absorbing them into graphs with 3-vertices only. That this is possible is due to the fact, as noted earlier, that the colour factors entering the 4-gluon vertex amount to products of colour factors entering 3-vertices. Furthermore, the power of the coupling in the 4-vertex is the same as the power of the coupling obtained by taking two 3-vertices (i.e. g^2). These facts are not coincidences but are ultimately a consequence of gauge invariance, as it was this that fixed the non-linear form of the gluon field strength tensor in Eq. (2.90).

Earlier, we have presented the Feynman rules for Yang–Mills theory and hopefully clarified their use through the example of $2 \to 2$ scattering of gluons. Let us now see how these remarks generalise to the case of an arbitrary amplitude, with m external particles and L loops. In order to construct such an amplitude, one would start by drawing all possible Feynman diagrams subject to the constraints of having a fixed m and L. This implies a certain overall power of the coupling, which we write explicitly in the following. Next, one can systematically eliminate all graphs with 4-gluon vertices by absorbing their contributions into different graphs containing *only* 3-vertices, such that the relevant colour factors match. One is left with a sum over effective diagrams analogous to those in Figure 3.8, where each diagram has a unique colour factor given by dressing all cubic vertices in a clockwise way with structure constants f^{abc}. The denominator factors in each graph are also fixed, as they arise from the propagator factors associated with each internal line. Finally, one must integrate over the L loop momenta $\{k_l\}$, such that one is led to the following general expression:

$$
\mathcal{A}_m^{(L)} = g^{m-2+2L} \sum_i \left(\prod_{l=1}^{L} \int \frac{d^4 p_l}{(2\pi)^4} \right) \frac{1}{S_i} \frac{c_i n_i}{\prod_{\alpha_i} p_{\alpha_i}^2}. \tag{3.23}
$$

The overall power of the coupling is indeed what one finds upon requiring m external particles and L legs and will be the same for each diagram. The sum is over all cubic effective diagrams i, where there are L loop momenta to be integrated over for each diagram, given that every diagram has the same loop number. We denote by $\{\alpha_i\}$ the set of internal lines, such that each has 4-momentum $p_{\alpha_i}^\mu$. Each diagram then has an appropriate colour factor c_i and a kinematic numerator n_i, where the latter contains all the residual dependence on momenta and polarisation vectors that is not explicitly shown in Eq. (3.23). There is also a symmetry factor S_i for each graph, which we briefly mentioned earlier when talking about Feynman rules. Again, we need not worry about this for our purposes.

Of course, Eq. (3.23) is not immediately useful but merely a general schematic form that can be surmised from the Feynman rules. All we have actually done is to translate the problem of calculating scattering amplitudes into that of finding the set of kinematic numerators $\{n_i\}$ for a given m and L. Furthermore, these numerators are defined at the level of the integrand, such that one must also still carry out the integrals over the loop momenta in order to obtain practical results. Nevertheless, Eq. (3.23) turns out to be a very useful way to think about the amplitude, as we now start to explore.

3.3   BCJ Duality

The kinematic numerators in Eq. (3.23) are not unique. For example, Feynman rules depend on which gauge we are working in. Thus, by performing a gauge transformation, we change the result of individual (effective) diagrams, such that the total result for the amplitude remains the same. There is also a greater freedom than this: we are clearly free to shift all the numerators in Eq. (3.23) according to

$$n_i \to n_i + \Delta_i, \tag{3.24}$$

provided that the set of shifts $\{\Delta_i\}$ satisfies

$$\sum_i \frac{c_i \Delta_i}{\prod_{\alpha_i} p_{\alpha_i}^2} = 0. \tag{3.25}$$

These transformations were first considered in Ref. [2] and referred to as *generalised gauge transformations*. It is then natural to ponder whether there is some special choice of the $\{n_i\}$ that one can make that may be particularly useful. Indeed there is, and to introduce the relevant ideas, let us return to the relatively simple case of $2 \to 2$ scattering, such that Eq. (3.23) reduces to Eq. (3.22). Furthermore, Eq. (3.25) simplifies considerably to

$$\frac{c_s \Delta_s}{s} + \frac{c_t \Delta_t}{t} + \frac{c_u \Delta_u}{u} = c_t \left(\frac{\Delta_s}{s} + \frac{\Delta_t}{t} \right) + c_u \left(\frac{\Delta_s}{s} + \frac{\Delta_u}{u} \right), \tag{3.26}$$

where on the right-hand side we have used Eq. (3.21) to express the shift constraint in terms of fully independent colour factors. This in turn means that the coefficient of each colour factor must vanish, implying

$$\frac{\Delta_t}{t} = \frac{\Delta_u}{u} = -\frac{\Delta_s}{s} \equiv \Delta,$$

where Δ is a common function of kinematic arguments (i.e. momenta and polarisations). We thus find that the general form of a generalised gauge transformation for the $2 \to 2$ scattering amplitude is

$$n_s \to n_s + s\Delta, \quad n_t \to n_t - t\Delta, \quad n_u \to n_u - u\Delta. \tag{3.27}$$

Let us denote the transformed numerators by $\{n_i'\}$, and let us next note that one may use the freedom of Eq. (3.27) to demand the following condition:

$$n'_s - n'_t - n'_u = 0. \tag{3.28}$$

To see this, consider a set of numerators $\{n_i\}$ that does not obey this condition. From Eq. (3.27), the latter then amounts to

$$n_s - n_t - n_u + (s + t + u)\Delta = 0 \;\Rightarrow\; \Delta = \frac{n_t + n_u - n_s}{s + t + u}. \tag{3.29}$$

We have thus succeeded in finding a shift parameter Δ such that Eq. (3.28) is obeyed.[3]

To see why Eq. (3.28) is interesting, note that it has precisely the same form as the Jacobi identity for the colour factors of the effective diagrams in Figure 3.8, namely Eq. (3.21). We know that the latter is a consequence of the fact that there is a local gauge symmetry underlying the theory, with an associated Lie group and Lie algebra. Equation (3.28) is in terms of completely different quantities – the kinematic numerators that enter the expressions for individual effective graphs. These have nothing to do with colour charge degrees of freedom and are instead functions of momenta and polarisation vectors. However, the fact that a similar equation arises implies – if we are very optimistic – that there might be some sort of symmetry algebra underlying the $\{n_i\}$. This suggestion was made in Ref. [2] and is known as *Bern–Carrasco–Johansson (BCJ) duality*. Here the word "duality" implies some sort of linking between the colour and "kinematic" Lie algebras, such that a Jacobi identity in one implies a Jacobi identity in the other.

To make such a bold claim, of course, we need a lot more evidence than Eq. (3.28)! In order to see how this works, let us see how the Jacobi identity of Eq. (3.21) generalises to effective diagrams that have more external legs and/or loops. At a given loop order, we may always choose to isolate a certain part of a given diagram and ignore the rest, as shown in Figure 3.9. Any given internal line will form

---

[3]Experts may complain here that the denominator of Eq. (3.29) vanishes for massless particles, which may be proven to be a consequence of momentum conservation. However, as discussed in Ref. [2], the condition of Eq. (3.28) actually turns out to hold in *any* gauge. Thus, Δ is well defined after all, albeit arbitrary. This special situation is limited to $2 \to 2$ scattering and does not affect our general arguments.

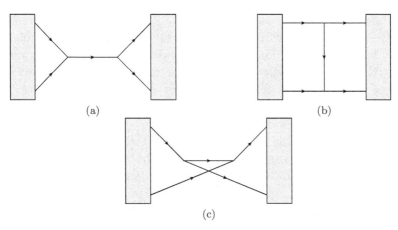

(a)   (b)

(c)

Fig. 3.9. Diagrams at higher-loop orders can be put into overlapping sets of three, where the diagrams in each set are identical up to a single section that is (a) s-, (b) t- or (c) u-channel like.

either an s-, a t-, or a u-channel-like configuration joined to the rest of the diagram, given that these are the only possibilities. Let us take the case where an s-channel-like diagram occurs, as in part (a) of the figure. We can then make two other diagrams by keeping the rest of the diagram the same but changing the middle of the diagram to a t- or u-configuration, as shown in Figures 3.9(b) and (c). These will be some other diagrams in the complete set of 3-vertex diagrams at this order, given that the number of loops and external lines will remain the same. Also, the colour factors of these three diagrams are related by a Jacobi identity. To see this, recall that the colour factor of any graph is found by dressing every 3-vertex in a clockwise manner with a set of structure constants. It follows that the colour factors of the diagrams in Figure 3.9 are all given by the product of a part which is the same for all the diagrams and a part corresponding to just the middle section we have isolated. The middle sections are related according to Eq. (3.21), using similar arguments to those used earlier. Thus, we must have

$$c_a = c_b + c_c \qquad (3.30)$$

for the total colour factors of the diagrams in Figure 3.9.

Applying the above arguments to all internal lines of all diagrams in the complete set of 3-vertex effective diagrams for a given m and L, it follows that this set can always be divided into (possibly overlapping) sets of three, such that the colour factors in each set are related by a Jacobi identity. The *BCJ duality conjecture* is then the statement that the kinematic numerators for the graphs can be similarly related, considered as functions of their momenta. I have used the word *conjecture* to describe this statement, as it has at the time of writing not been proven to hold for arbitrary numbers of loops (see Refs. [37, 38] for some of the most valiant attempts). Although we have considered the case of pure Yang–Mills theory here, extensions of the theory to include additional particle content and/or symmetries (e.g. supersymmetry) are also known to obey BCJ duality, in highly non-trivial cases up to four-loop order [3, 4, 7, 10, 39–66], depending on the theory.

If correct, BCJ duality is a remarkable property of non-abelian gauge theories that our traditional ways of thinking had completely obscured. This is because in most (generalised) gauge choices, the kinematic numerators will not obey Jacobi identities, such that a transformation as in Eq. (3.25) is needed in order to put the $\{n_i\}$ in *BCJ-dual form*. From a conceptual point of view, it is utterly strange that there should be a kinematic symmetry that somehow mirrors the colour Lie algebra. In the previous chapter, we used the analogy of actors moving on a stage when we talked about particles (the actors) moving in spacetime (the stage). In this analogy, kinematic properties refer to the movements of the actors on the stage. To represent (colour) charges, we could dress the actors in different costumes, depending on their charge. What BCJ duality is then telling us is that the possible movements of the actors on the stage are restricted by the type of costume they are wearing. For actors, this is perhaps not so strange: someone will have written a play that indeed tells us how people in certain costumes must behave. In the case of non-abelian gauge theory, however, we did not even know that the play existed, let alone what the story is!

In order to fully understand what BCJ duality is trying to tell us, we would need to know precisely what the kinematic symmetry algebra is for a given theory, whose Lie algebra implies the Jacobi-like identities for the kinematic numerators $\{n_i\}$. This is a perfectly sensible question, whose answer has remained elusive. There are very few cases in which we know what the kinematic algebra is, and they typically correspond to certain limits or restrictions of theories. We return to this topic in Chapter 6.

The discovery of BCJ duality is already interesting enough and remains an active research topic worldwide. One of its uses, for example, is to provide an alternative way to calculate amplitudes in gauge theory. At a given loop order, one can make an ansatz for the kinematic numerators $\{n_i\}$ in Eq. (3.23) e.g. an expansion in loop and external momenta with undetermined parameters. One can then impose BCJ duality by requiring that all kinematic Jacobi relations hold, and thus hope to fix the ansatz completely. If so, there are standard methods for checking that a given amplitude is correct, in that it passes all physical consistency checks (see e.g. Ref. [26] for a review). As well as its potential usefulness in gauge theory alone, BCJ duality goes hand-in-hand with an equally surprising phenomenon, which links together different types of field theories.

3.4 Gravity Amplitudes and the Double Copy

Equation (3.23) represents the general form of an amplitude in Yang–Mills theory and its generalisations. We can also consider scattering amplitudes in gravity theories. Historically, this was first investigated some decades ago by various authors (see e.g. Refs. [67–75]), who were trying to understand gravitational physics from a high-energy physics point of view. In gauge theory, the vacuum consists of zero field, and thus we can think of a non-zero field as a deviation from the vacuum state. The basic idea in quantum gravity is to consider the metric tensor $g_{\mu\nu}(x)$ as a "field" in spacetime, but we cannot take this to be zero in the vacuum. Rather, in the absence of gravity, we expect $g_{\mu\nu}$ to reduce to the Minkowski (flat) space metric $\eta_{\mu\nu}$. This motivates the decomposition

$$g_{\mu\nu}(x) = \eta_{\mu\nu} + \kappa h_{\mu\nu}, \qquad (3.31)$$

where the factor of

$$\kappa = \sqrt{16\pi G_N} \qquad (3.32)$$

is conventional, and we refer to $h_{\mu\nu}$ as the *graviton field*. It is zero in a vacuum. Furthermore, substituting Eq. (3.31) into the Einstein equations and keeping terms at linear order in κ only, one finds the gravitational wave solutions referred to in Chapter 1. Quanta of the field $h_{\mu\nu}$ are then hypothetical graviton particles of quantum gravity. More generally, one may expand the full non-linear Einstein equations as a series in κ, which generates an infinite series of terms involving progressively higher powers in $h_{\mu\nu}$, with indices contracted in various ways. As in Yang–Mills theory, these non-linear terms can be interpreted as interactions of multiple gravitons. One may also set up the language of Feynman rules, and thus one will find an infinite number of multi-graviton vertices, with new ones turning on at each order in perturbation theory: from Eq. (3.31), each power of the graviton field involves a power of κ.

Other than this, the Feynman rules for quantum gravity are similar in spirit to those in the previous section for Yang–Mills theory. In order to list the propagator and vertex factors, one must impose extra conditions on the graviton field, analogous to how one must fix a gauge in Yang–Mills theory. In gravity, this redundancy is associated with the fact that the metric is only defined up to coordinate transformations. We will not bother with listing explicit graviton Feynman rules but merely comment on their complexity. Unlike the 3-vertex in Yang–Mills theory, which from Eq. (3.8) has six distinct mathematical terms, the three-graviton vertex (in the most commonly used "gauge") has over 170! This makes calculations in quantum gravity stupendously unwieldy, such that even computers struggle after a while. There is also no obvious relationship between calculations in (non-abelian) gauge theories and those in gravity. Although the ethos of Feynman rules may be the same, the details of the calculation are sufficiently dissimilar as to confirm the fact that these are very different theories, describing very different physics.

This situation changed dramatically in Refs. [3, 4], which proposed a new way to calculate gravity amplitudes. One may start with a non-abelian gauge theory amplitude as in Eq. (3.23), where the numerators $\{n_i\}$ are understood to be in BCJ-dual form (i.e. manifestly obeying kinematic Jacobi identities). Then one may replace the coupling constant as follows:

$$g \rightarrow \frac{\kappa}{2}. \tag{3.33}$$

Physically, this corresponds to replacing the strength of the Yang–Mills force with that of gravity, in some appropriately normalised way. Next, we can strip off the colour factors $\{c_i\}$ from Eq. (3.23) and replace them with a second set $\{\tilde{n}_i\}$ of kinematic factors. The reason for adding a tilde is that these numerators may come from a *different* gauge theory than the original numerators $\{n_i\}$, a point that we elaborate on in the following. Finally, one obtains the following formula:

$$\mathcal{M}_m^{(L)} = \left(\frac{\kappa}{2}\right)^{m-2+2L} \sum_i \left(\prod_{l=1}^{L} \int \frac{d^4 p_l}{(2\pi)^4}\right) \frac{1}{S_i} \frac{n_i \tilde{n}_i}{\prod_{\alpha_i} p_{\alpha_i}^2}. \qquad (3.34)$$

It is not completely non-obvious to any practitioner of traditional QFT that this is a gravitational amplitude with L loops and m external legs. Certainly, it *might* be one: we have after all introduced the gravitational coupling constant κ and also removed all traces of colour charge, which gravity knows nothing about. That this procedure indeed yields gravity amplitudes has been checked in highly non-trivial cases and is also inspired by previous work [1], as we see in the following. It is called the *double copy*, where this refers to the fact that two copies of gauge theory numerators are taken in Eq. (3.34). You may often see the colloquial phrase that the double copy implies that gravity is somehow "the square of gauge theory". Whilst this helps us to remember Eq. (3.34), it is not literally true: the copy is taken term-by-term in the sum over 3-vertex diagrams, and only the numerator factors are copied. In particular, the denominators are left unchanged. Once a gravity amplitude has been generated using the double copy, one may check using standard methods (as for gauge theory) that it satisfies all known physical constraints and hence is correct.

Earlier, we have not been very precise about what we mean by a "gravity amplitude". The reason for this is that the double copy applies in a variety of gravity theories, depending on which two gauge theories we take the kinematic numerators $\{n_i\}$, $\{\tilde{n}_i\}$ from. We saw in Chapter 1, for example, that it is possible to add extra symmetries called *supersymmetries* in either gauge or gravity theories. These correspond to adding extra fermionic fields, such that each individual supersymmetry relates bosonic and fermionic degrees of freedom. It is common to denote by $\mathcal{N}$ the number of distinct supersymmetries in a given theory, and the maximum amount of supersymmetry one can add in four-dimensional gauge and gravity theory is $\mathcal{N} = 4$ and $\mathcal{N} = 8$, respectively. Taking both sets of gauge theory numerators in Eq. (3.34) from $\mathcal{N} = 4$ Super-Yang–Mills (SYM) theory gives an amplitude in $\mathcal{N} = 8$ supergravity. Likewise, taking numerators from $\mathcal{N} = 2$ and $\mathcal{N} = 4$ SYM gives $\mathcal{N} = 6$ supergravity, and so on. The key idea is that the number of individual polarisation states of particles has to match up on both sides of the double-copy correspondence. This in turn implies that the double copy of pure Yang–Mills theory with itself is not pure general relativity. To see this, note that the gluon has two polarisation states. Upon taking two copies, we then expect there to be *four* polarisation states in the gravity theory. In four spacetime dimensions, these turn out to be the graviton (which has two polarisation states) plus two additional scalar (spinless) fields. These are called the *axion* and the *dilaton*, and we see these again in what follows.

Like the existence of BCJ duality, the double copy is a conjecture, although it has been tested up to four-loop order in various theories [3, 7, 10, 47–66]. Higher-loop tests are possible in theories with more supersymmetry, as this turns out to reduce the number of non-zero numerators, making it easier to solve the BCJ duality ansatz. It turns out that in certain kinematic regions (i.e. particular limits of the real or virtual particle momenta), one can obtain amplitudes to all orders in perturbation theory. One such region is when any emitted radiation – either real or virtual – is *soft*, meaning that it has a low 4-momentum. Another interesting region is when the centre of mass energy is much greater than the momentum transfer between the incoming particles. This is called the *Regge limit*, and in both this and the soft limits, one may show that the double copy works to all orders in perturbation theory [76–83].

More significantly, for amplitudes with no loops $(L = 0)$, the double copy can be proven to work for arbitrary numbers m of external particles, where it turns out to be a consequence of previous work in string theory. We review the relevant ideas in the following section.

3.5　String Theory and the KLT Relations

We have repeated many times the statement that quantum field theory is a type of theory that successfully unifies quantum mechanics and special relativity. In doing so, it provides a fundamental

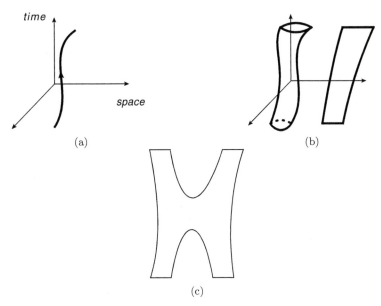

Fig. 3.10. (a) A particle traces out a worldline in spacetime as it moves; (b) strings trace out *worldsheets*, and examples are shown for both closed and open strings; (c) an open-string Feynman diagram.

explanation for the existence of particles, namely that they are excitations of wave-like configurations of fields. However, QFT is not the only theory that combines quantum theory with relativity. Another possibility is *string theory*, which differs from QFT in that the underlying objects are not point particles but one-dimensional strings instead.

To see the difference, note that a particle is a zero-dimensional object, in that it has no finite size in any spatial dimension. As is implicit when we draw Feynman diagrams, particles trace out *worldlines* as they move throughout spacetime, such that at each fixed time, we can locate the position of the particle, as in Figure 3.10(a). By contrast, a string is one-dimensional, meaning that it is defined by a curve in spacetime, such that we need a single parameter to tell us which "point" of the string we might be talking about. Given that curves may be closed or open, we also have *closed* or *open strings*. As these objects move throughout spacetime, they trace out a *worldsheet*, which is itself two-dimensional. Examples are shown for free strings in Figure 3.10(b), for both types of strings.

Point particles can interact in scattering processes, which in QFT are described by Feynman diagrams. Similarly, strings can also interact, and we can also draw Feynman diagrams for them. An example string Feynman diagram is shown in Figure 3.10(c), which shows the scattering of two open strings, with another two open strings in the final state. It is known how to write down the theory of quantum relativistic strings, and thus how to convert diagrams such as those shown in Figure 3.10(c) into scattering amplitudes (see e.g. Refs. [84–87] for textbook treatments). At low energies – which, from the uncertainty principle, corresponds to "zooming out" from the diagrams in Figures 3.10(b) and 3.10(c) – we cannot resolve the finite size of the string, and thus string theory must reduce to a quantum field theory of particles, as is indeed known to be the case.

Interestingly, string theory was originally written down to try to describe strong interactions, although a mixture of theoretical reasoning and experimental results showed that this was not the whole story. Indeed, we now know that QCD describes the strong interaction at current collider energies, such that the effective energy scale at which string theory deviates from QFT must be much higher. In the 1980s, string theory was resurrected as a theory that unifies gravity and the other interactions. If correct, this tells us that the energy scale at which string-like behaviour emerges is much closer to the *Planck scale* at which quantum gravity effects become important. This is too high to be able to probe directly in experiments, but it may well be that calculable effects in string theory left imprints in the early universe that we would see as fluctuations in the cosmic microwave background today.

To date, string theory is the only known theory that completely unifies non-abelian gauge theories and gravity into a single consistent mathematical framework. The price we pay, however, is that it is not at all clear how the particular QFTs relevant to our universe (e.g. the Standard Model of Particle Physics) emerge from string theory in a concrete way. Original hopes that our familiar theories would be uniquely fixed by string theory remain unfulfilled: the theory is only mathematically consistent in 10 spacetime dimensions, such that a mechanism is needed for curling up six of these extra dimensions so that we don't see them. This creates an enormous number of possible low-energy solutions of the theory, with no clear guidance on how to choose between them. It is also true that we lack a proper non-perturbative definition of the theory, and it is almost certainly true that our current way of thinking about string theory obscures some deep underlying structural features.

Even if string theory remains somewhat speculative as a fundamental theory of nature, it is highly useful as a mathematical tool for connecting different field theories that we know *are* relevant for the universe we live in. For our purposes, we note that Ref. [1] showed that certain scattering amplitudes for closed strings can be written as sums of products of scattering amplitudes for open strings. This set of correspondence has become known as the *KLT relations* and is important for the following reason. In the low-energy limit, closed strings turn out to correspond to gravitons, axions, or dilatons. Open strings, on the other hand, lead to non-abelian gauge fields (e.g. gluons) at low energies. Thus, the KLT relations in string theory imply some sort of relationship between scattering amplitudes in non-abelian gauge and gravity (field) theories. This relationship is precisely the double copy! Furthermore, the argument works for any number of particles in the gauge or gravity theory. The original form of this correspondence is for the case of maximally supersymmetric theories, such that closed and open strings yield $\mathcal{N} = 8$ supergravity and $\mathcal{N} = 4$ SYM theory, respectively [48]. However, the explanation will also work with less or no supersymmetry, so certainly applies to GR, minimally coupled to an axion and dilaton.

The KLT relations are very useful but alas are only set up for Feynman diagrams with no loops. Thus, whilst the KLT relations are compelling as an explanation for where the double copy comes from, they do not confirm the fact that, in field theory, the double copy appears to work at arbitrary loop level. It may be that this structure is present in field theory but not in string theory. Alternatively, it may be possible to generalise the KLT relations in string theory to arbitrary loop orders, although it is not at all clear how to do this at present (see Ref. [88] for progress at one-loop order). In the approximation for which the KLT argument for the double copy is valid, however, it should be stressed that it is not at all necessary for string theory to be a *bona fide* theory of nature. Rather, it acts as a mathematical bridge that makes structures manifest in physically relevant theories that were previously hidden. Indeed, the double copy was itself directly inspired by the KLT relations and would likely not have been found without them. This more than justifies the existence of string theory, which in this particular use has no need to be compared with experiment.

3.6 Biadjoint Scalar Theory and the Zeroth Copy

In going from Eq. (3.23) to Eq. (3.34), we stripped off the colour factors in the gauge theory and replaced them with a second set of kinematic numerator factors. We could equally well have done the opposite, removing all kinematic information in the numerator of each graph, in favour of a second set of colour factors $\tilde{c}_i$ associated with a potentially *different* Lie algebra to the first one. The resulting formula is

$$\mathcal{A}_m^{(L)} = y^{m-2+2L} \sum_i \left(\prod_{l=1}^{L} \int \frac{d^4 p_l}{(2\pi)^4} \right) \frac{1}{S_i} \frac{c_i \tilde{c}_i}{\prod_{\alpha_i} p_{\alpha_i}^2}, \tag{3.35}$$

where we have also relabelled the coupling constant. It is natural to consider this replacement and to then ask whether or not it constitutes a scattering amplitude in some theory. The answer is indeed yes, and the relevant theory is that of a spinless ("scalar") field with two different types of colour charges in general. Colour indices are also known as *adjoint indices* in the literature, and thus this theory is known as *biadjoint scalar (field) theory*. Its vacuum field equation is

$$\partial^2 \Phi^{aa'} + y f^{abc} \tilde{f}^{a'b'c'} \Phi^{bb'} \Phi^{cc'} = 0, \tag{3.36}$$

where $\{f^{abc}\}$ and $\{\tilde{f}^{a'b'c'}\}$ are structure constants associated with the two (potentially different) colour algebras, and the different types of colour index are distinguished by putting primes on one set of indices. The field $\Phi^{aa'}$ then indeed carries two types of colour charges, as required.

We know that biadjoint scalar theory is not a physically relevant theory by itself. Indeed, the energy of the theory is not bounded from below, making it difficult to interpret what the vacuum of the theory would be. However, an increasing amount of research has shown that at least some of the dynamics of biadjoint scalar theory is inherited by gauge and gravity theories, such that it is worth understanding in more detail. The process of converting the gauge theory amplitude of Eq. (3.23) into the biadjoint amplitude of Eq. (3.35) is known as the *zeroth copy*, and the inverse of this is commonly referred to as the *inverse zeroth copy*. Both of these processes are again expected to hold to arbitrary loop levels.

We can summarise the results of this chapter by saying that scattering amplitudes in three different types of theories can be written in a common form, using Eqs. (3.23), (3.34), and (3.35), such that simple procedures exist for converting between them. The full set of correspondences is shown in Figure 1.2 and offers a tantalising glimpse of some sort of hidden structure underlying all of the theories that are currently relevant for fundamental physics! Quite how widely we should interpret Figure 1.2 is itself unclear, and this very question will occupy us for the rest of this book.

Chapter 4

The Classical Double Copy: A First Look

In the previous chapter, we saw that scattering amplitudes in quantum field theory provide a concrete realisation of the scheme of Figure 1.2, in which quantities in different types of theories are mutually related. However, it is tempting to ponder whether Figure 1.2 can be interpreted much more generally than this. Does it, for example, apply to the complete theories, whatever this might mean? If so, we ought to be able to take any sort of object in one of the theories and match it with counterparts in one of the other theories. Note that we do not necessarily have to consider quantum properties but can also ask whether we can map features of the classical theories. For both gauge theory and gravity, a large number of solutions of the equations of motion are known, in perturbation theory or otherwise. We can then ask if it is possible to take particular solutions in gauge theory and exactly map them to counterparts in gravity, such that this procedure is directly related to the double copy for scattering amplitudes. Remarkably, the answer to this question is yes, and a number of cases are even known in which *exact* solutions can be mapped in this manner. Recent years have seen an increasing understanding of when this is possible, as well as how to formulate double copies of classical solutions in perturbation theory.

In this chapter, we introduce this topic by studying the first known example of a double copy for (exact) classical solutions, which was formulated in a double-copy context in Ref. [89] (see also Refs. [90, 91]

for earlier related work in a different context). It relies on a special class of solutions of general relativity, which we examine in the following section.

4.1   Kerr–Schild Solutions in GR

By a solution of general relativity, we mean an explicit form of the metric tensor $g_{\mu\nu}(x)$ as a function of spacetime that solves the Einstein equation of Eq. (2.105) (or Eq. (2.106) if there is a non-zero cosmological constant). This describes the local structure of spacetime at all possible spatial positions and times and thus in principle constitutes a complete history of a possible universe. In practice, a given solution for the metric may approach the Minkowski metric of flat space away from some localised region, such that one may instead consider a given solution to approximate the spacetime near some astrophysical object. Famous examples of solutions in general relativity include black holes and cosmological models for how our universe evolved during and after the Big Bang. Our own universe is considerably more complicated than any of these simplified models, but it can reasonably be hoped that they provide a decent approximation on very large scales, upon which our universe appears homogeneous. Note that one way to present a given solution for the metric is to simply quote the components of the metric $g_{\mu\nu}(x)$ as a function of the spacetime coordinates. A rather more compact way is to consider the line element

$$ds^2 = g_{\mu\nu}(x)dx^\mu dx^\nu \tag{4.1}$$

in a particular coordinate system. This represents the (squared) "length" of an infinitesimal displacement dx^μ in spacetime at the position x^μ, as measured using the metric tensor. For example, the line element for Minkowski space in Cartesian coordinates (t, x, y, z) is

$$ds^2 = -dt^2 + dx^2 + dy^2 + dz^2, \tag{4.2}$$

whereas in spherical polar coordinates (t, r, θ, ϕ), it is

$$ds^2 = -dt^2 + dr^2 + r^2 d\theta^2 + r^2 \sin^2 \theta d\phi^2. \tag{4.3}$$

Einstein's equations are a set of coupled non-linear partial differential equations and extremely complicated to solve in general. It is perhaps remarkable that exact solutions can be found at all, yet decades of ingenuity have established a growing catalogue of exact results: see e.g. [92, 93] for highly useful compendia. Typically, exact solutions possess some amount of symmetry or other special property, which can be used to formulate a suitable guess for the form of $g_{\mu\nu}(x)$, to be substituted as a trial solution into the Einstein equation. One such guess is known as the *Kerr–Schild ansatz*, which first arose in Refs. [94, 95] (see Ref. [92] for an excellent modern review). It consists of writing the metric according to Eq. (3.31), where $h_{\mu\nu}$ takes the following form:

$$h_{\mu\nu}(x) = \phi(x)k_\mu(x)k_\nu(x). \tag{4.4}$$

We see that the graviton is taken to decompose into a so-called *outer product* of a (position-dependent) 4-vector k_μ, which is then multiplied by an additional scalar field. Furthermore, the vector $k_\mu(x)$ cannot be arbitrary but must instead satisfy certain conditions. First, we have

$$\eta^{\mu\nu}k_\mu k_\nu = g^{\mu\nu}k_\mu k_\nu = 0, \tag{4.5}$$

such that the graviton is null with respect to both the Minkowski and full metrics. This in turn implies that the inverse metric is given by

$$g^{\mu\nu} = \eta^{\mu\nu} - \frac{\kappa}{2}\phi(x)k^\mu(x)k^\nu(x). \tag{4.6}$$

Next, we have the so-called *geodesic condition*

$$k \cdot \partial k^\mu \equiv k^\nu \partial_\nu k^\mu = 0, \tag{4.7}$$

which can be given a geometric interpretation that we will explore in later chapters. It is likely not at all obvious to you why the earlier ansatz is useful. However, substituting Eq. (4.4) into Eq. (2.105) and using the conditions of Eqs. (4.5) and (4.7) leads to a wondrous simplification: the Einstein equations become *linear*, if thought about in the right way! That is, if we consider the Ricci tensor with one index upstairs and the other downstairs, it takes the form

$$R^\mu{}_\nu = \frac{\kappa}{2}\left(\partial^\mu \partial_\alpha(\phi k^\alpha k_\nu) + \partial_\nu \partial^\alpha(\phi k_\alpha k^\mu) - \partial^2(\phi k^\mu k_\nu)\right). \tag{4.8}$$

Care is needed in interpreting this equation. On the left-hand side, we have raised the index using the full metric, as is right and proper. On the right-hand side, however, indices have been raised with the Minkowski metric only, and the exactness of this result may be confirmed by explicit calculation (warning: this takes a few pages more algebra than you might think!). Importantly, nothing has been neglected, such that there are no corrections to Eq. (4.8) that are higher order in κ.

Looking at Eq. (4.8), we see that all terms on the right-hand side indeed contain only a single power of the graviton, upon using Eq. (4.4). Thus Eq. (2.105) has indeed become a linear equation, which has two advantages. First, it is much easier to solve for explicit forms of $\phi(x)$ and $k_\mu(x)$, a given choice of which constitutes a distinct Kerr–Schild solution (up to the possibility that two Kerr–Schild solutions may be related by a coordinate transformation and thus physically equivalent). Second, once we have found a solution, we know that it is exact, with no additional corrections.

4.2   The Kerr–Schild Double Copy

As we have seen, the double copy relates properties in biadjoint scalar, gauge, and gravity theory, according to the scheme of Figure 1.2. We now show that the special form of Kerr–Schild solutions indeed allows us to take single and zeroth copies, which we later understand are intimately related to the double copy for scattering amplitudes. We consider *static solutions*, namely those that do not explicitly depend on time:

$$\partial_0\phi = \partial_0 k_\mu = 0. \tag{4.9}$$

We will also set $k^0 = 1$ in Eq. (4.4), absorbing any non-trivial dynamics into the field $\phi(x)$. One then finds that specific components of the Ricci tensor of Eq. (4.8) are given by

$$R^0{}_0 = \frac{\kappa}{2}\nabla^2\phi, \quad R^i{}_0 = -\frac{\kappa}{2}\partial_j\left[\partial^i(\phi k^j) - \partial^j(\phi k^i)\right]. \tag{4.10}$$

Next, consider the Einstein equations in the vacuum case (i.e. with energy–momentum tensor $T_{\mu\nu} = 0$). Equation (2.105) becomes

$$R_{\mu\nu} - \frac{R}{2} g_{\mu\nu} = 0, \tag{4.11}$$

such that contracting the indices implies

$$R = 0 \quad \Rightarrow \quad R_{\mu\nu} = 0.$$

Hence, for vacuum Kerr–Schild solutions, the results of Eq. (4.10) can be combined into the single equation

$$\partial_\mu \left[\partial^\mu(\phi k^\nu) - \partial^\nu(\phi k^\mu) \right] = 0. \tag{4.12}$$

Comparison with Eqs. (2.26) and (2.30) shows that this resembles one of the Maxwell equations, and we can make this a lot more concrete as follows. Given a Kerr–Schild solution as in Eq. (3.31), we may define a gauge field according to

$$A_\mu^a = g c^a \phi(x) k_\mu(x), \tag{4.13}$$

where g is the gauge theory coupling constant and c^a a constant colour vector. Upon substituting this into the expression for the Yang–Mills field strength (Eq. (2.90)), this becomes

$$F_{\mu\nu}^a = g c^a \left[\partial_\mu(\phi k_\nu) - \partial_\nu(\phi k_\mu) \right] - g^2 f^{abc} c^b c^c \phi^2 k_\mu k_\nu.$$

The second term vanishes, given that f^{abc} is antisymmetric in its indices, but $c^b c^c$ is symmetric. This leaves the simpler result

$$F_{\mu\nu}^a = g c^a \left[\partial_\mu(\phi k_\nu) - \partial_\nu(\phi k_\mu) \right], \tag{4.14}$$

which when substituted into the vacuum Yang–Mills field equation (Eq. (2.91) with $J_\nu^a = 0$) yields

$$\partial^\nu F_{\mu\nu}^a = 0. \tag{4.15}$$

We thus find that, for every static vacuum Kerr–Schild solution, there is a corresponding gauge theory solution given by Eq. (4.13). Furthermore, this solution linearises the Yang–Mills equations, such that they essentially reduce to the Maxwell equations[1] but where the gauge field contains an arbitrary colour vector c^a. This is perhaps unsurprising, given that the field equation in gravity is itself linearised by the Kerr–Schild ansatz.

By analogy with Figure 1.2, we call the gauge field of Eq. (4.13) the *(Kerr–Schild) single copy* of the corresponding gravity solution, and it is simply obtained via the replacements

$$\frac{\kappa}{2} \to g, \quad k_\mu \to c^a. \tag{4.16}$$

That is, the standard gravitational coupling constant that appears in the scattering amplitude of Eq. (3.34) is replaced by its gauge theory counterpart, and the Kerr–Schild vector k_μ is replaced with a colour vector. This process can clearly be repeated, and by doing so we obtain the scalar field

$$\Phi^{aa'} = y c^a \tilde{c}^a \phi, \tag{4.17}$$

where we now have two colour vectors $(c^a, \tilde{c}^{a'})$, associated with two potentially different colour (Lie) algebras, and have also replaced the gauge theory coupling with its biadjoint counterpart. Equation (4.17) has the right properties to be a biadjoint field. Substituting it into the field equation of Eq. (3.36) shows that this is indeed satisfied and also linearises:

$$\partial^2 \Phi^{aa'} = 0. \tag{4.18}$$

Following Figure 1.2, we call this the *(Kerr–Schild) zeroth copy* of the given gravity solution, and we have thus obtained the following result:

Every stationary vacuum Kerr–Schild solution in general relativity has a corresponding single-copy gauge field, and zeroth copy

---

[1]Above, we have examined only one of the Maxwell equations. However, the Bianchi identity of Eq. (2.31) will also be satisfied, given that the Yang–Mills field strength reduces to a linearised version.

biadjoint field. Each solution can be chosen to linearise its respective field equation.[2]

We have thus seen our first example of how the double copy can be extended to classical solutions. To complete this story, however, we must explain why the above set of results relates to the double copy for scattering amplitudes that we described in the previous chapter. A full explanation of why it is indeed equivalent will need the techniques developed in the following chapters. However, we can already give some preliminary remarks here. First, note that both the amplitude and Kerr–Schild double copies involve the replacement of kinematic (position, momentum, or spin-dependent) information in gravity with (colour) charge information in the gauge theory. In the amplitude case, this is the replacement of kinematic numerators by colour factors. In the Kerr–Schild case, it is the replacement of the vector field $k_\mu(x)$ by a colour vector. Next, note that in both double copies, there are quantities that are left untouched by the copying procedure. For amplitudes, these are the denominators associated with each cubic graph, and that can be thought of as being composed of propagator factors, as if these originate from a pure scalar theory. In the Kerr–Schild case, it is the field $\phi(x)$ that is left untouched. However, this is the solution of a scalar theory and thus formally similar to the amplitudes case. Another similarity we can mention at this point is that the earlier argument is insensitive to the number of spacetime dimensions. Thus, we expect the Kerr–Schild double copy to work in arbitrary higher dimensions $d > 4$, a property which is shared with the double copy for amplitudes [3, 4].

4.3   Examples

The earlier discussion has been rather abstract, due to the fact that we have presented a general argument valid for any stationary Kerr–Schild solution. Let us clarify this by considering particular examples, including those that go beyond the simple assumptions made earlier.

---

[2]Our reason for the somewhat careful wording about "choosing" solutions to linearise the field equations is due to the fact that one can always perform a gauge transformation that makes them non-linear.

4.3.1   *The Schwarzschild black hole*

Arguably the most famous solution of general relativity – and the first to be found historically – is the *Schwarzschild black hole*. It is the unique[3] physical object one finds by searching for a spherically symmetric vacuum solution of Eq. (2.105) and is famous for its *event horizon*, defining a region of space from which not even light can escape, hence the name "black hole". The form of the metric depends, as usual, on the particular coordinate system chosen. Crucially for our purposes, the Schwarzschild solution has a Kerr–Schild form, where the relevant graviton field is given by Eq. (4.4) with

$$\phi = \frac{M}{4\pi r}, \quad k_\mu = (1,1,0,0), \tag{4.19}$$

where we have used spherical polar coordinates (t, r, θ, ϕ). According to Eq. (4.13), we may take a single copy of this result. Contracting also with the generators $\mathbf{T}^a$, we may write the corresponding gauge field as

$$\mathbf{A}_\mu = \frac{g\mathbf{c}}{4\pi r}(1,1,0,0), \quad \mathbf{c} = \sum_a c^a \mathbf{T}^a, \tag{4.20}$$

where we have also relabelled the mass M by absorbing it into c^a. Equation (4.20) will not necessarily look familiar, but one may perform a gauge transformation to put this into the following form:

$$\mathbf{A}_\mu = \left(\frac{g\mathbf{c}}{4\pi r}, 0, 0, 0 \right). \tag{4.21}$$

As stated in Eq. (2.25), the first component of the gauge field A^μ represents the electrostatic potential so that

$$A_\mu = (-V, \boldsymbol{A}).$$

Thus, our single copy of the Schwarzschild black hole corresponds to an abelian-like gauge theory solution with electrostatic and magnetic vector potential

$$V = -\frac{g\mathbf{c}}{4\pi r}, \quad \boldsymbol{A} = 0, \tag{4.22}$$

---

[3]The fact that the Schwarzschild solution is the only possible spherically symmetric object is guaranteed by a result known as *Birkhoff's theorem*.

respectively. We can recognise this as directly analogous to the potentials for a point charge in electromagnetism. A general point charge in the latter theory will have some charge Qe, where e is the magnitude of the charge on the electron and Q the amount of charge measured in these units. The factor appearing in Eq. (4.22) is similar: the coupling constant g plays the role of e, and the factor c specifies how much of each type of colour charge we have, in units of g. The minus sign is not physically meaningful, as we can always choose to absorb this in our definition of c. From Eq. (4.17), we may also take the zeroth copy of our solution, which in component form reads

$$\Phi^{aa'} = \frac{y c^a \tilde{c}^{a'}}{4\pi r}. \tag{4.23}$$

Again, we can interpret this as some sort of point-like particle sitting at the origin, but now with two different types of colour charges.

In this, our first example of a classical double copy, we see that the effect of the single copy is to replace mass in the gravity theory with charge in the gauge theory. The latter is replaced by the appropriate charge in biadjoint theory when performing the zeroth copy, and to further make sense of this behaviour, we can consider the sources needed to produce the earlier solutions. Although we have referred to the Schwarzschild solution as a vacuum solution with zero energy–momentum tensor, this is not strictly true. Rather, the energy–momentum tensor must be zero everywhere outside a spherically symmetric region containing the mass M, and the minimal source we can take is a point-like mass at the origin. The corresponding energy–momentum tensor is known to be

$$T^{\mu\nu} = M v^\mu v^\nu \delta^{(3)}(\boldsymbol{x}), \quad v^\mu = (1, 0, 0, 0), \tag{4.24}$$

where the three-dimensional delta function localises the particle to the origin as required, and the vector v^μ points in the time direction. Upon carefully substituting the single-copy field of Eq. (4.13) into the linearised Yang–Mills equations, one finds a source current

$$\mathbf{j}^\mu = -g c v^\mu \delta^{(3)}(\boldsymbol{x}), \tag{4.25}$$

which is indeed that for a static point charge at the origin. A similar (scalar) source can be derived in the biadjoint theory. Furthermore, the earlier analysis generalises straightforwardly to the *Tangherlini black hole*, which is the analogue of the Schwarzschild black hole in higher dimensions.

4.3.2   *The Kerr black hole*

After the Schwarzschild black hole, the next most complicated example is the *Kerr black hole*, which is rotating and chargeless. It has the following Kerr–Schild form:

$$\phi = \frac{\kappa}{2}\frac{1}{4\pi}\frac{Mr^3}{r^4 + a^2 z^2}, \quad k_\mu = \left(1, \frac{rx + ay}{r^2 + a^2}, \frac{ry - ax}{r^2 + a^2}, \frac{z}{r}\right), \qquad (4.26)$$

where we have used Cartesian coordinates (t, x, y, z) and introduced a *spheroidal radius* defined implicitly via

$$\frac{x^2 + y^2}{r^2 + a^2} + \frac{z^2}{r^2} = 1 \qquad (4.27)$$

except for the disc-like region in the (x, y) plane given by $\{x^2 + y^2 \leq a^2, z = 0\}$, where r is defined to be zero. Looking at Eq. (4.26) tells us that the z-direction has been picked out as special, and indeed it coincides with the axis of rotation of the black hole. Furthermore, a is related to the angular momentum of the black hole, as can be seen by the fact that we recover the Schwarzschild metric in the limit $a \to 0$. As for Schwarzschild, the Kerr solution is vacuum everywhere outside a certain region, and the minimal source one can consider is that of a rotating disc of mass of radius a in the (x, y) plane. To make this more concrete, one can introduce the *spheroidal coordinates* (t, r, θ, ϕ), where the spatial coordinates are given in terms of Cartesians via

$$x = \sqrt{r^2 + a^2}\sin\theta\cos\phi, \quad y = \sqrt{r^2 + a^2}\sin\theta\sin\phi, \quad z = r\cos\theta,$$
$$(4.28)$$

in terms of which the minimal energy–momentum tensor is found to be [89, 96, 97]

$$T^{\mu\nu} = \frac{\kappa}{2}\delta(z)\Theta(a - \rho)\left(-\frac{M\sec^3\theta}{4\pi a^2}\right)[\xi^\mu\xi^\nu - \cos^2\theta\,\tilde\eta^{\mu\nu}]. \qquad (4.29)$$

Here we have defined the 4-vector (in the spheroidal coordinate system)

$$\xi^\mu = \left(1, 0, \frac{1}{a}, 0\right), \tag{4.30}$$

as well as the tensor (in the Cartesian coordinate system (t, x, y, z))

$$\tilde{\eta}^{\mu\nu} = \text{diag}(-1, 1, 1, 0). \tag{4.31}$$

Furthermore, in Eq. (4.29), we have used the Heaviside (step) function

$$\Theta(x) = \begin{cases} 1, & x \geq 0, \\ 0, & x < 0. \end{cases} \tag{4.32}$$

The combination of the delta and Heaviside functions in Eq. (4.29) then tells us that the source is only non-zero on the disc of radius r in the (x, y) plane. The physical interpretation of the two distinct terms is that they constitute a rotating mass profile and a radial pressure on the disc itself [89, 96].

Upon forming the single copy according to the earlier procedure, one finds that it solves the abelian-like Yang–Mills equations, with a source current

$$j^\mu = q\xi^\mu, \quad q = \delta(z)\Theta(a - \rho)\frac{gc}{4\pi a^2}\sec^3\theta. \tag{4.33}$$

As in the Schwarzschild case, the mass has been replaced with (colour) charge, where the overall profile of charge is the same as the mass profile in the gravity case. Upon comparing Eqs. (4.29) and (4.33), however, we see that whilst the first term in the gravity source appears to itself be some sort of "double copy" of the gauge theory current, the second (radial pressure) term has no counterpart. Physically, this is unsurprising: like charges repel in gauge theory, but (like) masses attract in gravity. Thus, in one of the theories, a radial pressure is needed to stabilise the rotating disc so that it remains stable and produces a static solution. Interestingly, then, the sources need not formally double copy in order that the fields do. Quite how general sources behave under the double copy is an open question (see e.g. Ref. [98] for a pioneering study in this area) but may be particularly important for generalising the classical double copy yet further.

The above algebra is all very fine but not particularly illuminating. To bring the single copy of the Kerr solution – informally known as √Kerr in the literature – to life, we can plot the electric and magnetic fields **E** and **B**, given that the single copy is abelian-like. The electric field is shown in Figure 4.1, where the rotating disc of charge is seen sideways on. At short distances, there is a non-trivial profile to the field, due to the particular profile of charge

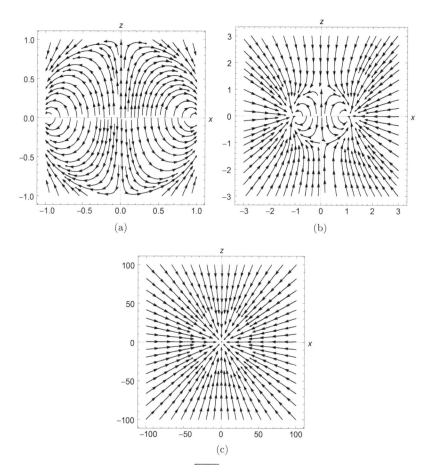

Fig. 4.1.   The electric field of the √Kerr solution, seen at (a) short; (b) medium and (c) long distance. The rotating disc of charge is seen sideways on from $-1 \leq x \leq 1$.

on the disc (which happens to diverge on the ring at the disc's edge). As we zoom out, the structure of the disc is no longer visible, and the field thus becomes that of a point charge. Similarly, the magnetic field is shown in Figure 4.2. A complicated structure at short distances gives way to a magnetic dipole field at large distances. The reason for the latter is that a rotating disc of charge looks like a set of nested current loops, which are indeed associated with a dipole field. As discussed in Ref. [89], the Kerr–Schild double

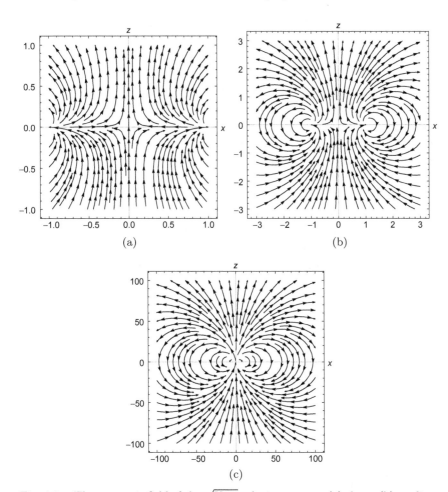

Fig. 4.2.   The magnetic field of the $\sqrt{\text{Kerr}}$ solution, seen at (a) short; (b) medium and (c) long distance. The rotating disc of charge is seen sideways on from $-1 \leq x \leq 1$.

copy can also be applied to higher-dimensional generalisations of the Kerr black hole, specifically the well-known family of *Myers–Perry solutions* [99].

4.3.3 (Anti-)de Sitter space

As discussed in Chapter 2, one may consider adding a cosmological constant term in Einstein's field equation, leading to Eq. (2.106) rather than Eq. (2.105). The vacuum solution of this equation is known as *de Sitter (dS)* spacetime if $\Lambda > 0$, and *Anti-de Sitter (AdS)* spacetime if $\Lambda < 0$. These solutions have widespread uses throughout both theoretical and applied physics. The chief motivation for studying de Sitter spacetime is that the cosmological constant in our own universe is indeed found to be positive, albeit very small. Furthermore, AdS spacetime is of direct relevance for the so-called *AdS/CFT correspondence* that relates strongly coupled gauge theory to a weakly coupled string theory in a higher-dimensional AdS space [100]. This has led to new methods of analysing strongly coupled systems, with applications ranging from heavy ion collisions to condensed matter physics.

We may choose to rewrite Eq. (2.106) as an Einstein equation with no cosmological constant, Eq. (2.105), but with the cosmological constant term appearing on the right-hand side, and interpreted as a non-zero energy–momentum tensor. If we are solving the vacuum equation, then this will be a deviation about Minkowski space, such that the energy–momentum tensor turns out to be

$$T_{\mu\nu} = -\left(\frac{\kappa^2}{2}\right)^{-1} \Lambda \eta_{\mu\nu}. \qquad (4.34)$$

The (A)dS solution can then be chosen to have a Kerr–Schild form, with

$$\phi = \frac{\kappa}{2}(6\Lambda r^2), \quad k_\mu = (1,1,0,0), \qquad (4.35)$$

where we have used spherical polar coordinates (t, r, θ, ϕ). The single-copy gauge field is

$$\mathbf{A}^\mu = -g\mathbf{c}(6\rho r^2)k_\mu, \qquad (4.36)$$

where c is a constant colour vector as usual, and substitution into the Maxwell equations reveals that there is a constant charge density ρ filling all space [89]. This makes sense given that the cosmological constant in gravity is a uniform energy density: the single copy must turn this into a charge density, to be consistent with the previous examples (and with the kinematic/colour replacements for scattering amplitudes).

4.4 Beyond Single Kerr–Schild Solutions

The Kerr–Schild double copy constitutes a family of exact classical gravity solutions, which may be assigned counterparts in gauge and gravity theory. Each such solution will not have an explicit Kerr–Schild form if we transform to a different coordinate system in general. Conversely, it is also true that not all gravitational solutions can be cast into a Kerr–Schild form. Quite how special this class of solution is is explored in more detail in the following chapter. However, in the spirit of broadening the scheme of Figure 1.2 yet further, we can already ask whether there are exact double copies that go beyond the simple stationary Kerr–Schild construction outlined earlier. By now, a number of explicit examples are known.

4.4.1 *Double Kerr–Schild solutions*

One way of generalising the Kerr–Schild double copy is to assume a *double Kerr–Schild* form. That is, one may generalise Eq. (4.4) to

$$h_{\mu\nu} = \phi(x)k_\mu(x)k_\nu(x) + \psi(x)l_\mu(x)l_\nu(x), \qquad (4.37)$$

where $\phi(x)$ and $\psi(x)$ are two independent scalar fields, and the vectors k_μ and l_μ satisfy

$$k^2 = l^2 = k \cdot l = 0, \quad k \cdot \partial k_\mu = l \cdot \partial l_\mu = 0. \qquad (4.38)$$

As in the single Kerr–Schild case, all contractions in this equation can be understood as occurring with either the Minkowski, or the full metric. Whilst straightforward to write down, Eq. (4.37) is not a simple extension of the Kerr–Schild case. In particular, it does not linearise the expression for the Ricci $R^\mu{}_\nu$ tensor in general. However, for

certain very special cases, then a linear form is indeed achieved. One
may then form a single-copy gauge field by generalising Eq. (4.13):

$$A_\mu^a = c^a g \phi k_\mu + d^a \tilde{g} \psi l_\mu, \tag{4.39}$$

where we have assumed a different colour vector in the second
term and also relabelled the coupling constant to $\tilde{g}$, for reasons
that will become clear in the following. This gauge field then solves
the abelian-like gauge theory field equation which, given the latter is a linear equation, applies to each term separately. Note
that the single copy is taken term-by-term in the gravity solution,
which is analogous to how the single copy for scattering amplitudes
operates.

One particular example of a double Kerr–Schild metric is the
Taub–NUT solution [101, 102]. This consists of a point-like object
at the origin which has a Schwarzschild-like mass M but also a
NUT charge N which gives rise to a non-trivial rotational charac-
ter to the gravitational field that does not die off at infinity. One
may also include the effect of non-zero angular momentum (as in
the Kerr black hole), such that the source becomes a disc in the
(x, y) plane. This is then known as the *Kerr–Taub–NUT solution*.
As for the single Kerr–Schild metrics considered earlier, the double
Kerr–Schild form will not be manifest in arbitrary coordinate systems. Unfortunately, in the Taub–NUT case, the coordinates in which
the metric has the appropriate form are particularly complicated
and non-intuitive. First, it is known that in the so-called *Plebanski
coordinates*

$$p \in [-a, a], \quad q \in [0, \infty], \quad \sigma \in [0, 2\pi/a], \quad \tau \in [-\infty, \infty] \tag{4.40}$$

the line element for the Kerr–Taub–NUT solution – including also a
non-zero cosmological constant – is

$$ds^2 = \frac{q^2 - p^2}{\Delta_p} dp^2 + \frac{q^2 - p^2}{\Delta_q} dq^2 - \frac{\Delta_p}{q^2 - p^2}(d\tau + q^2 d\sigma)^2$$

$$- \frac{\Delta_q}{q^2 - p^2}(d\tau + p^2 d\sigma)^2, \tag{4.41}$$

with

$$\Delta_p = \gamma - \epsilon p^2 + \lambda p^4 - 2Np, \quad \Delta_q = -\gamma + \epsilon q^2 - \lambda q^4 - 2Mq, \tag{4.42}$$

where M and N are the mass and NUT charge, ϵ is a constant, γ is related to the angular momentum, and λ to the cosmological constant [103]. Then, Ref. [104] showed that upon transforming to the new variables $\tilde{\sigma}$ and $\tilde{\tau}$ defined by

$$d\tilde{\tau} = d\tau + \frac{p^2 dp}{\Delta_p} - \frac{q^2 dq}{\Delta_q}, \quad d\tilde{\sigma} = d\sigma - \frac{dp}{\Delta_p} + \frac{dq}{\Delta_q}, \tag{4.43}$$

the metric indeed has a double Kerr–Schild form:

$$g_{\mu\nu} = \bar{g}_{\mu\nu} + \phi k_\mu k_\nu + \psi l_\mu l_\nu, \tag{4.44}$$

where the background line element (arising from $\bar{g}_{\mu\nu}$) is

$$d\bar{s}^2 = -\frac{1}{q^2 - p^2} \left[\bar{\Delta}_p (d\tilde{\tau} + q^2 d\tilde{\sigma})^2 + \bar{\Delta}_q (d\tilde{\tau} + p^2 d\tilde{\sigma})^2 \right]$$

$$- 2(d\tilde{\tau} + q^2 d\tilde{\sigma}) dp - 2(d\tilde{\tau} + p^2 d\sigma) dq, \tag{4.45}$$

where

$$\bar{\Delta}_p = \gamma - \epsilon p^2 + \lambda p^4, \quad \bar{\Delta}_q = -\gamma + \epsilon q^2 - \lambda q^4. \tag{4.46}$$

Explicit forms of the Kerr–Schild functions and 4-vectors in the $(\tilde{\tau}, \tilde{\sigma}, p, q)$ system are

$$k_\mu = (1, q^2, 0, 0), \quad l_\mu = (1, p^2, 0, 0), \quad \phi = \frac{2Mq}{q^2 - p^2}, \quad \psi = \frac{2Np}{q^2 - p^2}. \tag{4.47}$$

To make sense of this, we can transform to spheroidal coordinates (t, r, θ, ϕ) according to

$$\tau = t + a\phi, \quad \sigma = \frac{\phi}{a}, \quad q = r, \quad p = a\cos\theta, \tag{4.48}$$

where $a^2 = \gamma$, and where (τ, σ) are the original coordinates appearing in the line element of Eq. (4.41). Following this transformation, the background line element can be shown to reduce to the usual Minkowski form if the cosmological constant contribution λ

is taken to zero. Furthermore, the Kerr–Schild vectors end up as follows:

$$k_\mu = \left(1, \frac{\rho^2}{a^2 + r^2}, 0, -a\sin^2\theta\right), \quad l_\mu = \left(1, 0, \frac{i\rho^2\csc\theta}{a}, -\frac{a^2 + r^2}{a}\right),$$
(4.49)

where

$$\rho^2 = r^2 + a^2\cos^2\theta.$$
(4.50)

Let us now interpret the single copy of this result, as obtained from Eq. (4.39), in the limit in which the angular momentum is taken to be zero, $a \to 0$. It is not immediately clear that one can do this, given that the Kerr–Schild vector l_μ diverges. However, the Kerr–Schild functions expressed in spheroidal coordinates are

$$\phi = \frac{2Mr}{r^2 - a^2\cos^2\theta}, \quad \psi = \frac{2Na\cos\theta}{r^2 - a^2\cos^2\theta},$$
(4.51)

such that the product ψl_μ is indeed perfectly well defined as $a \to 0$. The only non-zero components of the electric and magnetic fields of the single copy are found to be [105][4]

$$\mathbf{E}_r = -\frac{g\mathbf{c}}{4\pi r^2}, \quad \mathbf{B}_r = -\frac{\tilde{g}\mathbf{d}}{4\pi r^2}.$$
(4.52)

This looks like a point-like electric colour charge $\mathbf{c}$ in units of g and a point-like magnetic charge $\mathbf{d}$ in units of $\tilde{g}$. It is thus a so-called *dyon*, namely a particle possessing both electric and magnetic charges. The magnetic charge considered by itself makes the particle a *magnetic monopole*. Such particles are taken not to exist in Maxwell's theory of electromagnetism but may well exist in the full non-abelian gauge theories underlying fundamental interactions in our universe. Thus far, searches for magnetic monopoles have remained negative, and their mass must be high to avoid cosmological implications. The fact that magnetic and electric charges are in principle different to each

---

[4]Reference [105] reports results for the field strength tensor, but given the abelian-like nature of the solution, this can easily be converted into the electric and magnetic fields.

other explains why we have used two coupling constants g and $\tilde{g}$ earlier.

We already saw in the Schwarzschild and Kerr examples that the single copy maps mass in the gravity theory to electric charge in the gauge theory. We now get to enlarge this dictionary: NUT charge in gravity maps to magnetic charge in gauge theory. Analogies between the NUT solution and gauge theory magnetic monopoles have been made before (see e.g. Ref. [106] for a pedagogical review of the NUT solution), typically at large values of the radial coordinate r. The single copy is much nicer than this: it exactly relates the magnetic monopole and NUT solutions, for all values of the spacetime coordinates. The trick, however, is to identify a coordinate system such that the NUT solution takes a Kerr–Schild form. Before moving on, it is worth pointing out that higher-dimensional generalisations of the NUT solution exist, such that there are

$$n = \left\lfloor \frac{D}{2} \right\rfloor - 1$$

generalised NUT charges in D spacetime dimensions, where $\lfloor x \rfloor$ denotes the integer part of x. Obtained in Refs. [104, 107], these solutions were shown to have a multiple Kerr–Schild form in Ref. [107]. They can thus be single-copied, where each NUT charge will map to a corresponding magnetic charge. The zeroth copy of the Taub–NUT solution is also straightforward: the earlier four-dimensional example yields a biadjoint field

$$\Phi^{aa'} = c^a \tilde{c}^{a'} y\phi + d^a \tilde{d}^{a'} \tilde{y}\psi, \tag{4.53}$$

where we have allowed for two potentially different coupling factors y and $\tilde{y}$. As in the gauge theory case, we can interpret each term as being due to a different type of charge possessed by a point-like particle.

4.4.2 *Non-stationary solutions*

Another extension of the earlier construction is to consider non-stationary solutions, given that there are indeed examples of single Kerr–Schild solutions that depend explicitly on time [92]. Some preliminary examples are given in Ref. [89], which focused on arguably the simplest time-dependent solutions of gravity and gauge theory, namely wave-like solutions moving at the speed of light. Waves in both gauge theory and gravity are transverse, and we may consider the family of the so-called *pp-wave solutions* in gravity, whose wave-fronts are planes, but where the metric is not necessarily constant on each transverse plane. These have a Kerr–Schild form, which is most easily written by introducing the light-cone coordinates

$$u = z - t, \quad v = z + t, \tag{4.54}$$

in terms of Cartesian coordinates (t, x, y, z). Then the Minkowski line element takes the form

$$ds^2 = dudv + dx^2 + dy^2, \tag{4.55}$$

and the Kerr–Schild form of a pp-wave moving in the positive z-direction takes the form

$$\phi \equiv \phi(u, x, y), \quad k_\mu = (1, 0, 0, 0), \tag{4.56}$$

where k_μ is expressed in the (u, v, x, y) coordinate system. We see that the profile of the wave depends only on time through the combination $u = z - t$, which indeed represents a wave-like disturbance moving in the $+z$ direction at the speed of light (n.b. we are in natural units, with $c \to 1$). The function ϕ cannot be arbitrary but must satisfy the constraint

$$\left(\frac{\partial^2}{\partial x^2} + \frac{\partial^2}{\partial y^2} \right) \phi = 0, \tag{4.57}$$

which arises after substituting the earlier solution into the Einstein equations. A similar solution can be obtained for a wave travelling in the $-z$ direction, by interchanging the roles of u and v. Also, by rotating our original Cartesian coordinate system, we can generate a plane wave solution moving in any direction we choose.

The existence of plane wave solutions in gauge theories is also well known. In the abelian case, these are the original electromagnetic wave solutions first derived by Maxwell. Non-abelian plane waves were first derived in Ref. [108]. They are fully non-linear solutions in general, but a gauge may be chosen so that they obey the linearised equations. This matches what one expects upon taking a Kerr–Schild single copy of pp-waves: generalising the stationary Kerr–Schild procedure to the present case, one may define a gauge theory plane wave

$$A_\mu^a = c^a \phi k_\mu, \tag{4.58}$$

with ϕ and k_μ given as in Eq. (4.56) (up to relabellings of couplings, and mass parameters by charge parameters). This is indeed found to both linearise and solve the Yang–Mills equations, and thus formalises the fact that the non-abelian plane waves of Ref. [108] are indeed a single copy of gravitational pp-waves.[5] One may also take the zeroth copy, to give a biadjoint wave solution

$$\Phi^{aa'} = c^a \tilde{c}^{a'} \phi. \tag{4.59}$$

A particularly well-known example of a gravitational pp-wave is a *shockwave*, obtained by taking a stationary massive particle described by the Schwarzschild solution and infinitely boosting it whilst rescaling parameters to keep the total energy finite. This procedure was first carried out by Aichelburg and Sexl in Ref. [110], and the solution takes the earlier *pp*-wave form, with

$$\phi(u, x, y) = C\delta(u) \log |x^2 + y^2|, \tag{4.60}$$

for some constant C. The profile in the transverse plane is shown in Figure 4.3. It diverges as one moves towards the origin of the plane, due to the fact that the metric will diverge on the worldline of the boosted particle. The interpretation of the gauge and biadjoint solutions is similar: the field corresponds to that of an infinitely boosted charged particle, where rescalings have ensured that the total energy is finite. For an explicit discussion of how the boosting procedure can be carried out in each theory, making the double copy manifest throughout, see Ref. [111].

In subsequent chapters, we explore in detail why the Kerr–Schild double-copy procedure is directly related to the original double copy for scattering amplitudes. In the context of shockwaves, however, it is worth already mentioning an interesting point of contact. It is

---

[5]For an earlier discussion of the similarities between gauge and gravity plane waves, see Ref. [109].

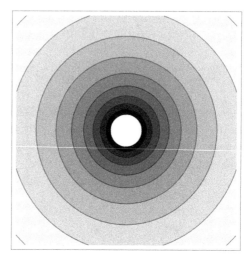

Fig. 4.3. Profile of the Aichelburg–Sexl shockwave solution in the transverse plane.

possible to obtain the Aichelburg–Sexl metric in gravity by summing up the results of Feynman diagrams to all orders in perturbation theory in the high-energy limit. As shown in Ref. [80], one may also carry out this procedure in gauge theory, such that one explicitly identifies the Aichelburg–Sexl metric as an amplitude-based double copy of a gauge theory shockwave!

What pp-waves have in common with the stationary solutions considered earlier is that the solution for the fields does not depend explicitly on at least one of the coordinates of the spacetime: stationary solutions do not depend explicitly on t, and pp-waves in the $+z$ direction do not depend on v. Thus, there is a *symmetry* of the solution under translations of this coordinate and a corresponding vector – called a *Killing vector* – that points in this direction. Reference [112] provided a more formal framework for generalising the stationary Kerr–Schild construction. Starting with the linearised Einstein equation for a Kerr–Schild metric, the authors considered contracting this with a Killing vector and showed that under certain circumstances the linearised gauge theory equations can be obtained, analogous to how Eq. (4.12) emerges in the stationary case. This provides a general framework which can accommodate both the stationary and wave-like solutions described earlier. A more general time-dependent solution was considered in Ref. [113], namely that corresponding to an accelerating black hole. This radiates energy in the form of gravitational waves, and its single copy is a radiating accelerating charge whose gauge field is related to the well-known *Liénard–Wiechert potential* in electromagnetism.

Before closing this section, it is worthwhile noting that the Kerr–Schild approach has been extended to the framework of *double field theory (DFT)*, a novel field theory with a doubled set of spacetime coordinates, which arises due to its various connections with string theory (see e.g. Ref. [114] for a review). In particular, Ref. [115] formulated the Kerr–Schild double copy in this framework, showing that solutions could be accommodated which have the full set of degrees of freedom arising from the double copy of pure Yang–Mills theory i.e. the axion and dilaton, in addition to the graviton. Extensions to more supersymmetric theories are considered in Ref. [116]. In Ref. [117], the DFT approach was used to construct relations for generalisations of amplitudes in which one of the particles remains virtual/off-shell [117] that generalise the field theory KLT relations discussed in Section 3.5.

4.5   The Kerr–Schild Double Copy in Curved Space

So far, we have considered defining the graviton field as the deviation from the Minkowski metric $\eta_{\mu\nu}$. There is another possibility, however, namely that we can take a non-trivial background metric $\bar{g}_{\mu\nu}$, such that the graviton is defined via

$$g_{\mu\nu} = \bar{g}_{\mu\nu} + \kappa h_{\mu\nu}, \tag{4.61}$$

where $g_{\mu\nu}$ is the full metric. This is the approach that is taken when considering small perturbations around a given solution, which might be e.g. a black hole or a cosmological model. Respective applications would include looking at quantum aspects of black hole physics or looking at the structure of perturbations in the cosmic microwave background.

Given Eq. (4.61), we may generalise the definition of a Kerr–Schild solution to curved space as follows (see e.g. Ref. [92] for a review). A Kerr–Schild graviton field can still be written in the form of Eq. (4.4), but where the null and geodesic conditions now amount to

$$g^{\mu\nu}k_\mu k_\nu = \bar{g}^{\mu\nu}k_\mu k_\nu, \quad k^\mu D_\mu k_\nu = 0, \tag{4.62}$$

where D_μ is the covariant derivative associated with the background metric $\bar{g}_{\mu\nu}$. The Ricci tensor with index placement as in Eq. (4.8) becomes

$$R^\mu{}_\nu = \bar{R}^\mu_\nu - \kappa h^\mu{}_\rho \bar{R}^\rho_\nu + \frac{\kappa}{2} D_\rho(D_\nu h^{\mu\rho} + D^\mu h^\rho{}_\nu - D^\rho h^\mu{}_\nu), \tag{4.63}$$

where $\bar{R}^\mu_\nu$ is the Ricci tensor associated with the background metric. Thus, it remains linear in the graviton field. In considering the single copy of a given gravity solution, there are two possibilities. First, there is a so-called *type A* single copy, in which we look for a gauge theory solution in Minkowski space, of the form

$$A^a_\mu = \bar{A}^a_\mu + \tilde{A}^a_\mu. \tag{4.64}$$

Here $\bar{A}^a_\mu$ is a background field that is itself a single copy of the background gravity solution $\bar{g}_{\mu\nu}$ and $\tilde{A}^a_\mu$ a perturbation that is a single copy of the curved-space graviton $h_{\mu\nu}$. Likewise, one may take a zeroth copy, defining a biadjoint field

$$\Phi^{aa'} = \bar{\Phi}^{aa'} + \tilde{\Phi}^{aa'}, \tag{4.65}$$

where $(\bar{\Phi}^{aa'}, \tilde{\Phi}^{aa'})$ are the zeroth copies of $(\bar{A}^a_\mu, \tilde{A}^a_\mu)$, respectively. A similar scheme has been proposed in the context of scattering amplitudes, in e.g. Refs. [118–122], and thus this procedure is expected to be the "correct" classical equivalent of the double copy in curved spacetimes.

However, there is also an interesting second possibility, namely that one can look for a single-copy gauge field that lives in the curved background $\bar{g}_{\mu\nu}$, which is now considered to be non-dynamical. That is, the gauge field must obey the *curved space* Yang–Mills equation

$$D^\mu F^a_{\mu\nu} = 0, \tag{4.66}$$

and this procedure was called a *type B double copy* in Ref. [123]. A similar procedure was also studied in Ref. [112], and a number of non-trivial examples were found, including black hole solutions built

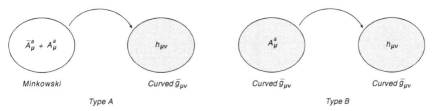

Type A   Type B

Fig. 4.4. In the type A double copy, a graviton perturbation is related to corresponding perturbations in gauge and biadjoint theory, where the background fields also double copy; in the type B double copy, the graviton is related to a gauge field living in the *same* curved background metric.

upon (A)dS background metrics. Furthermore, Ref. [123] argued that the type B double copy should hold for arbitrary *conformally flat* metrics i.e. those that can be written as

$$\bar{g}_{\mu\nu} = \Omega^2(x)\eta_{\mu\nu} \tag{4.67}$$

for some function $\Omega^2(x)$. It is not always possible to interpret the zeroth copy, however, and it is not expected that the type B procedure is fully general: in the original amplitude double copy, the gauge theory is manifestly in flat space. Nevertheless, the fact that the type B curved-space double copy includes solutions of astrophysical relevance (e.g. black holes and cosmological spacetimes) suggests that it may yet be useful for something. We summarise the two double-copy procedures in Figure 4.4.

In this chapter, we have described how the double copy that relates scattering amplitudes in biadjoint, gauge, and gravity theories can be generalised to certain types of exact classical solutions. At this stage, however, there are many unanswered questions. What entitles us, for example, to fully claim that the Kerr–Schild and scattering amplitude procedures are different facets of the *same* underlying correspondence? How special is the class of exact solutions that we are able to double copy, and can it be generalised further? Furthermore, why does the double copy for the classical solutions considered earlier involve products in position space, whereas the double copy for amplitudes involves products in momentum space? We will be able to answer all of these questions in the following chapter, where we will first take a detour to examine another way in which one can formulate the exact classical double copy.

Chapter 5

Exact Classical Double Copies

In the previous chapter, we have seen our first example of a double copy relating exact solutions of biadjoint, gauge, and gravity theories, namely the Kerr–Schild double copy of Ref. [89]. Since then, a number of other double-copy procedures have appeared, with the aim of relating exact classical solutions. The aim of this chapter is to review the most widely studied examples and also to point out how they relate to each other. The first one we study is called the *Weyl double copy* and was first presented in Ref. [124]. In order to describe this procedure in more detail, we first need to develop the appropriate language, which is the subject of the following section.

5.1 The Spinorial Formalism

As described in Chapter 2, the equations describing the various field theories underlying fundamental physics are typically written in terms of 4-vectors and tensors, in order to make the requirements of special relativity manifest. Mathematically speaking, this is because such quantities are acted on straightforwardly by Lorentz transformations. An alternative formalism exists, however, in terms of quantities called spinors. Whilst it may look complicated to the uninitiated, the spinor language ends up being extremely efficient for performing certain calculations in general relativity, as well as for showing up general structures that would be much more cumbersome in the tensor approach. Spinors occur in a variety of contexts in physics, and there are different ways of introducing them. In the present context, a classic reference work is that of Ref. [125], which has directly inspired the following discussion.

5.1.1 *Spinors and their geometry*

Consider a null position vector

$$x^\mu = (t, x, y, z), \quad x^2 = -t^2 + \boldsymbol{x}^2 = 0. \tag{5.1}$$

This has four real degrees of freedom subject to a single constraint, and it is possible to parametrise the components using two complex numbers λ_0 and λ_1, such that the null condition is automatically satisfied:

$$t = \frac{1}{\sqrt{2}} \left(\lambda_0 \lambda_0^* + \lambda_1 \lambda_1^* \right),$$

$$x = \frac{1}{\sqrt{2}} \left(\lambda_0 \lambda_1^* + \lambda_1 \lambda_0^* \right),$$

$$y = \frac{1}{i\sqrt{2}} \left(\lambda_0 \lambda_1^* - \lambda_1 \lambda_0^* \right),$$

$$z = \frac{1}{\sqrt{2}} \left(\lambda_0 \lambda_0^* - \lambda_1 \lambda_1^* \right). \tag{5.2}$$

We can then collect together the earlier parameters into two two-component objects:

$$\lambda_A = (\lambda_0, \lambda_1), \quad \bar{\lambda}_{A'} = (\lambda_0^*, \lambda_1^*), \tag{5.3}$$

where the indices $A, A' \in \{0, 1\}$. The object λ_A is called a *spinor*, and its conjugate $\bar{\lambda}_{A'}$ is called a *dual spinor*. We note that it is conventional to place a prime on the index of a dual spinor, and our reason for regarding (dual) spinors as different types of objects is clarified in the following. To clarify the notation further, note that we would write the individual components of the earlier dual spinor as

$$\bar{\lambda}_{0'} = \lambda_0^*, \quad \bar{\lambda}_{1'} = \lambda_1^*.$$

Given these quantities, we can consider the outer product

$$x_{AA'} \equiv \lambda_A \bar{\lambda}_{A'} = \frac{1}{\sqrt{2}} \begin{pmatrix} t+z & x+iy \\ x-iy & t-z \end{pmatrix}, \tag{5.4}$$

where Eq. (5.2) has been used to write the various components in terms of the original position coordinates. The determinant of this matrix is

$$|x_{AA'}| = -\frac{1}{2}(-t^2 + x^2 + y^2 + z^2) = 0, \tag{5.5}$$

where we recognise the form of the Lorentz-invariant dot product for the position vector. Let us write $\mathbf{X}$ for the matrix whose components are $x_{AA'}$. Then, the determinant of $\mathbf{X}$ will be preserved by transformations of the form

$$\mathbf{X} \to \mathbf{A} \mathbf{X} \mathbf{A}^\dagger, \tag{5.6}$$

where

$$\mathbf{A} = \begin{pmatrix} \alpha & \beta \\ \gamma & \delta \end{pmatrix}, \quad |\mathbf{A}| = \alpha\delta - \beta\gamma = 1. \tag{5.7}$$

That is, $\mathbf{A}$ is a general 2×2 complex matrix with unit determinant. We may write the components of the transformed $\mathbf{X}$ matrix as

$$x'_{A'B'} = \frac{1}{\sqrt{2}} \begin{pmatrix} t'+z' & x'+iy' \\ x'-iy' & t'-z' \end{pmatrix},$$

and it follows that the parameters (t', x', y', z') will be superpositions of the original coordinates (t, x, y, z). Given that the determinant of the new matrix is similar to Eq. (5.5) but with primes on all coordinates, it follows that the transformation of Eq. (5.6) is associated with a Lorentz transformation of the position coordinates.[1] In index

---

[1]The explicit form of the Lorentz transformation matrix acting on $x^\mu = (t, x, y, z)$ in terms of the parameters appearing in Eq. (5.7) is cumbersome and can be found in Ref. [125].

notation, Eq. (5.6) implies the following separate transformations for the (dual) spinors entering Eq. (5.4):

$$\lambda_A \to A_A^B \lambda_B, \quad \bar{\lambda}_{A'} \to (A^\dagger)_{A'}^{B'} \bar{\lambda}_{B'}, \qquad (5.8)$$

where we denote the matrix components of $\mathbf{A}$ and $\mathbf{A}^\dagger$ by $A_A{}^B$ and $(A^\dagger)_{A'}{}^{B'}$, respectively. These are linear transformations, where the matrix in each case must have unit determinant. They are conventionally called *special linear transformations*, and the group of such transformations in this case is denoted by $\mathrm{SL}(2, \mathbb{C})$, which stands for "special linear 2×2 complex matrices" (n.b. this notation may be compared with that used for the $\mathrm{SU}(N)$ groups discussed in Chapter 2). From Eq. (5.8), we can see that spinors and dual spinors transform differently: if a spinor transforms according to a given matrix, its conjugate spinor transforms according to the conjugate transformation. This is our reason for regarding spinors and dual spinors as separate objects with different indices, albeit linked via complex conjugation. Note that there is not a one-to-one relationship between a given Lorentz transformation and its corresponding $\mathrm{SL}(2, \mathbb{C})$ transformation(s): we see from Eq. (5.6) that the matrices $\mathbf{A}$ and $(-\mathbf{A})$ will correspond to the same Lorentz transformation, given that the additional sign in the latter will cancel out. Thus, there is a two-to-one map from the set of $\mathrm{SL}(2, \mathbb{C})$ transformations to the set of possible Lorentz transformations on real spacetime coordinates.

The earlier discussion is extremely abstract, but there is in fact a way that one can visualise spinors. For the earlier null vector, let us define the dimensionless coordinates

$$\hat{x} = \frac{x}{r}, \quad \hat{y} = \frac{y}{r}, \quad \hat{z} = \frac{z}{r}, \qquad (5.9)$$

where

$$r^2 = x^2 + y^2 + z^2$$

and we must have $t = r$ from the null condition. The latter can then be written as

$$\hat{x}^2 + \hat{y}^2 + \hat{z}^2 = 1, \qquad (5.10)$$

which defines a sphere of unit radius, points of which correspond to the different null *directions* that can emanate from the origin. Given a

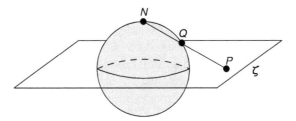

Fig. 5.1. Stereographic projection of a point Q on the unit sphere to a point P on the plane shown.

sphere, we may perform a so-called *stereographic projection* onto a plane that bisects it, as shown in Figure 5.1. A given point Q in the sphere can be associated with a point P on the plane by drawing a line from the north pole N of the sphere that passes through Q. Let us now associate the plane with the complex plane of a given complex variable ζ. We can choose to orientate the real and imaginary axes with the $\hat{x}$ and $\hat{y}$ directions, such that $\hat{z}$ is the vertical direction in the figure. Then, if the point Q is associated with coordinates $(\hat{x}, \hat{y}, \hat{z})$, then the point P can be shown to have complex position

$$\zeta = \frac{\hat{x} + i\hat{y}}{1 - \hat{z}}. \tag{5.11}$$

Taking the complex conjugate of this equation, one may then verify the relations

$$(\hat{x}, \hat{y}, \hat{z}) = \left(\frac{\zeta + \zeta^*}{1 + \zeta\zeta^*}, \frac{\zeta - \zeta^*}{i(1 + \zeta\zeta^*)}, \frac{\zeta\zeta^* - 1}{1 + \zeta\zeta^*} \right). \tag{5.12}$$

If we then write

$$\zeta = \frac{\lambda_0}{\lambda_1}, \tag{5.13}$$

we recover the parametrisation of Eq. (5.2). We thus see that the ratio of a given spinor's components directly determines a null direction. If we only care about the direction of a null vector, and not its length in a spacetime diagram, then we can fix $\lambda_1 = 1$, such that $\zeta = \lambda_0$ uniquely determines the direction. However, not all directions are obtainable by this construction: Eq. (5.13) can only be applied if $\lambda_1 \neq 0$. The case $\lambda_1 = 0$ corresponds to $\zeta \to \infty$, which corresponds to the north pole N of the sphere itself. We must thus add to the complex plane a separate "point at infinity", such that the complex plane of ζ is then referred to as the *extended complex plane*. Also, the sphere in this construction is referred to as a *Riemann sphere*, such that we see that the Riemann sphere has a one-to-one correspondence with the extended complex plane. In order to describe the north pole, we can instead use a stereographic projection from the south pole rather than the north pole so that the north pole appears as a well-defined point in the complex plane (the origin).

Given two spinors λ and ξ, it is convenient to define a product that is preserved under $\mathrm{SL}(2, \mathbb{C})$ transformations which, as described earlier, correspond to Lorentz transformations of the corresponding null vectors. The relevant product – known as an *inner product* – can be written as

$$\langle \lambda, \xi \rangle = \lambda_0 \xi_1 - \lambda_1 \xi_0, \tag{5.14}$$

and to show why, note that under Eq. (5.8) we have

$$\langle \lambda, \xi \rangle \to (\alpha\delta - \beta\gamma)\langle \lambda, \xi \rangle,$$

as may be verified by explicit calculation. By the determinant condition of Eq. (5.7) – itself a consequence of the fact that the dot product for 4-vectors has to be invariant – we see that this product is indeed preserved. In index notation, we may write Eq. (5.14) as

$$\langle \lambda, \xi \rangle = \epsilon^{AB} \lambda_A \xi_B, \tag{5.15}$$

where

$$\epsilon^{AB} = \begin{pmatrix} 0 & 1 \\ -1 & 0 \end{pmatrix} \tag{5.16}$$

is the two-dimensional Levi-Civita symbol. The fact that $\epsilon^{AB} = -\epsilon^{BA}$ means that, as can be directly observed in Eq. (5.14), the spinor product is antisymmetric:

$$\langle \lambda, \xi \rangle = -\langle \xi, \lambda \rangle.$$

By analogy with 4-vector algebra, we can introduce upstairs components of a spinor:

$$\lambda^A = \epsilon^{AB} \lambda_B, \tag{5.17}$$

such that the inner product between two spinors can be written as

$$\langle \lambda, \xi \rangle = \lambda^A \xi_A. \tag{5.18}$$

We thus see that the symbol ϵ^{AB} acts like the metric tensor for 4-vectors, allowing us to raise indices. Extreme care is needed, however. Unlike the metric tensor, which is symmetric in its indices, the Levi-Civita symbol is antisymmetric so that the precise ordering of indices is important. In particular, we have

$$\lambda_A \xi^A = \epsilon^{AB} \lambda_A \xi_B = -\epsilon^{BA} \lambda_A \xi_B = -\lambda^A \xi_A \neq \lambda^A \xi_A.$$

This is a bugbear of any theoretical physicist who has ever written a research document featuring spinors, including the author of this book! Similarly, we can lower indices as follows:

$$\lambda_A = \lambda^B \epsilon_{BA}, \quad \epsilon_{BA} = \begin{pmatrix} 0 & 1 \\ -1 & 0 \end{pmatrix}, \tag{5.19}$$

which we can check is indeed consistent with Eq. (5.17). The earlier relationships are summarised for spinors with unprimed indices. For dual spinors, we can introduce

$$\epsilon^{A'B'} = \begin{pmatrix} 0 & 1 \\ -1 & 0 \end{pmatrix}, \quad \epsilon_{A'B'} = \begin{pmatrix} 0 & 1 \\ -1 & 0 \end{pmatrix}, \tag{5.20}$$

such that

$$\bar{\lambda}^{A'} = \epsilon^{A'B'} \bar{\lambda}_{B'}, \quad \bar{\lambda}_{A'} = \bar{\lambda}^{B'} \epsilon_{B'A'}. \tag{5.21}$$

5.1.2 Tensors as spinors

In Eq. (5.4), we have seen that we can identify a null 4-vector in spacetime with a matrix in spinor space. It is common to refer to all quantities in spinor space as *(multi-index) spinors*, in the same way that 4-vectors and objects with more than one spacetime index are all referred to as *tensors*. Thus, we see that a null 4-vector is associated with a two-index spinor, and we may formalise this relationship by writing Eq. (5.4) in the form

$$x_{AA'} = x^\mu \sigma_{\mu \, AA'}, \tag{5.22}$$

where we have introduced the *Infeld–van der Waerden symbols*

$$\sigma_{\mu\,AA'} = \frac{1}{\sqrt{2}} \left\{ \begin{pmatrix} 1 & 0 \\ 0 & 1 \end{pmatrix}, \begin{pmatrix} 0 & 1 \\ 1 & 0 \end{pmatrix}, \begin{pmatrix} 0 & i \\ -i & 0 \end{pmatrix}, \begin{pmatrix} 1 & 0 \\ 0 & -1 \end{pmatrix} \right\}. \tag{5.23}$$

These quantities have a single spacetime index, and 2 spinor indices, thus constituting a 4-vector of matrices in spinor space. Implicit in the index labelling is that these symbols are not unique but instead transform appropriately under Lorentz transformations in spacetime and/or SL$(2, \mathbb{C})$ transformations in spinor space. In more pedestrian terms, we are free to choose a different coordinate system in either spacetime or spinor space, and this will change the components of the Infeld–van der Waerden symbols in a prescribed way. Similarly, spacetime or spinor indices can be raised or lowered using the metric tensor and/or Levi-Civita symbols, as described earlier.

As might already be clear, Eq. (5.22) is not limited to null vectors. Once we have the Infeld–van der Waerden symbols, we can convert *any* tensorial quantity into a multi-index spinor. The simplest extension is to an arbitrary 4-vector, for which we may define

$$V_{AA'} = V^\mu \sigma_{\mu\,AA'} = \frac{1}{\sqrt{2}} \begin{pmatrix} V^0 + V^3 & V^1 + iV^2 \\ V^1 - iV^2 & V^0 - V^3 \end{pmatrix}. \tag{5.24}$$

The determinant of this matrix is given by

$$|V_{AA'}| = -V_\mu V^\mu \tag{5.25}$$

so that the conclusion of Eq. (5.5) generalises to arbitrary time-like and space-like 4-vectors. By raising indices and contracting, we may also derive the useful relation

$$V_{AA'} W^{AA'} = -\frac{1}{2} V_\mu W^\mu. \tag{5.26}$$

More generally, any tensor with a number of upper and lower indices can be converted into a multi-index spinor by appropriate contraction with Infeld–van der Waerden symbols:

$$T_{A_1 A_1' A_2 A_2' \ldots B_1 B_1' B_2 B_2'} = T^{\alpha_1 \ldots \beta_1}{}_{\alpha_2 \ldots \beta_2}$$
$$\times \left(\sigma_{\alpha_1 A_1 A_1'} \cdots \sigma_{\beta_1 B_1 B_1'} \right) \left(\sigma^{\alpha_2}{}_{A_2 A_2'} \cdots \sigma^{\beta_2}{}_{B_2 B_2'} \right). \tag{5.27}$$

Due to the nature of the Infeld–van der Waerden symbols, every spacetime (tensor) index becomes a pair of spinor indices, one primed and one unprimed. Conversely, it is not true that every spinor has a tensorial counterpart: this clearly applies to any spinor with a non-equal number of (un)primed indices. Nevertheless, spinors with an odd number of indices of a single type play a role in nature, in describing particles of non-integer spin.[2]

So far we have seen that any tensor can be converted into a (multi-index) spinor. If this is the first time that you are seeing this formalism, then you will be forgiven for wondering what on earth the value of doing this is, given that tensors and 4-vectors are already complicated enough. The answer, as usual in physics, is that different languages can often make certain features much clearer. The spinor formalism in particular has two very nice properties, which allows us to greatly simplify how we talk about solutions of gauge theory and gravity. Let us take each of these properties in turn.

Reduction to symmetric spinors

A symmetric spinor is one whose components do not care about interchanging the order of indices. More formally, we may write this for a single-index type as

$$\psi_{AB...C} = \psi_{(AB...C)}. \qquad (5.28)$$

Here we have used a conventional notation for symmetrising over indices:

$$\psi_{(A_1 A_2...A_n)} = \frac{1}{n!} \sum_\pi \psi_{A_{\pi_1} A_{\pi_2}...A_{\pi_n}}, \qquad (5.29)$$

where the right-hand side contains a sum over all permutations π of $(1, 2, \ldots, n)$. A consequence of the fact that spinor indices can only take one of two values $(A, A' \in \{0, 1\})$ is that all multi-index spinors with a single-index type (primed/unprimed) can be reduced to sums of products of Levi-Civita symbols and symmetric spinors.

---

[2]As an example, the 4-component spinors entering the Dirac equation can be thought of as a combination of two 2-component spinors of the type described in this chapter.

To see this, let us first consider the simplest case of a multi-index spinor with two unprimed indices. On general grounds, we may write

$$\xi_{AB} = \xi_{(AB)} + \xi_{[AB]}, \qquad (5.30)$$

with

$$\xi_{(AB)} = \frac{1}{2}\left[\xi_{AB} + \xi_{AB}\right], \quad \xi_{[AB]} = \frac{1}{2}\left[\xi_{AB} - \xi_{AB}\right], \qquad (5.31)$$

i.e. we may decompose ξ_{AB} as a sum of its symmetric and antisymmetric parts. For the antisymmetric part, we may note that a general 2×2 antisymmetric matrix has the form

$$\begin{pmatrix} 0 & K \\ -K & 0 \end{pmatrix} \propto \epsilon_{AB}.$$

We may thus write

$$\xi_{AB} = \xi_{(AB)} + K\epsilon_{AB},$$

such that contracting with ϵ^{AB} on both sides yields

$$K = \frac{1}{2}\xi^{A}{}_{A} \quad \Rightarrow \quad \xi_{AB} = \xi_{(AB)} - \frac{1}{2}\xi^{C}{}_{C}\epsilon_{AB}.$$

Now considering spinors with higher numbers of indices, a similar argument yields

$$\eta_{...AB...} = \eta_{...(AB)...} - \frac{1}{2}\epsilon_{AB}\eta_{...}{}^{C}{}_{C...},$$

and this result may be used iteratively to remove all antisymmetric parts of a given spinor, leaving only symmetric spinors and Levi-Civita symbols, as asserted earlier. For more details, see e.g. Ref. [125].

Factorisation into principal spinors

The earlier decomposition of multi-index spinors into simpler (symmetric) building blocks is already nice, but in fact we may go further than this. Any symmetric spinor can itself be decomposed into a symmetrised product of single-index spinors:

$$\psi_{AB...C} = \alpha_{(A}\beta_B \cdots \gamma_{C)}, \qquad (5.32)$$

where the spinors appearing on the right-hand side are known as *principal spinors*. To see this, let us quote a beautifully simple argument from Ref. [126]. First, we introduce an arbitrary spinor

$$\xi^A = (x, y)$$

and form the quantity

$$\psi_{AB...C}\xi^A\xi^B \cdots \xi^C.$$

This is a polynomial of degree n in the complex quantities x and y and thus must be completely factorisable into the form

$$(\alpha_0 x - \alpha_1 y)(\beta_0 x - \beta_1 y) \ldots (\gamma_0 x - \gamma_1 y).$$

Each factor has the form of a spinor product $\lambda_A \xi^A$. Furthermore, the original multi-index spinor is symmetric in its indices so that we are unavoidably led to the form of Eq. (5.32).

The earlier properties already give a clue as to the power of the spinor formalism: complicated objects (multi-index spinors) can be reduced to combinations of much simpler objects (single-index spinors). This reveals a great deal of structure underlying field theories that is either invisible, or extremely cumbersome, in the traditional tensor approach. To see this in action, let us discuss what some of the theories of Chapter 2 look like if we use the spinor language.

5.2   Spinor Fields

We started this chapter by pointing out that null vectors emanating from a single point could be associated with a single-index spinor. Then, we saw that arbitrary tensors could also be mapped to a spinor, which will have multiple indices in general. Generalising this discussion yet further, we can consider tensors that are defined differently at different points in spacetime. This is precisely the definition of a *tensor field*, and we saw in Chapter 2 that such fields enter many of our theories of fundamental interactions, such as electromagnetism, Yang–Mills theory, and gravity. By contracting a given tensor field with Infeld–van der Waerden symbols, we get a *spinor field*, and thus all known field equations can be cast in terms of such objects. If we want to, we can think of having introduced an abstract "spin space" at each point in spacetime, in which our spinor objects live. To make this mathematically precise, we must then define rules for saying how the basis spinors of this spin space change as we move around our spacetime. This in turn amounts to saying how the Infeld–van der Waerden symbols change as we go from point to point, and we can ignore this complication in flat space, such that we may use the symbols in Eq. (5.23) everywhere.

Let us now take two canonical examples of tensor field theories, namely the vacuum equations for electromagnetism and gravity, and see what these look like in the spinorial formalism. For electromagnetism, the field equations are written in terms of the antisymmetric field strength tensor $F_{\mu\nu} = -F_{\nu\mu}$. Its spinorial translation will then have the form

$$F_{AA'BB'} = -F_{BB'AA'}. \tag{5.33}$$

Given that the relative ordering of unprimed and primed indices is unimportant, we can use Eq. (5.33) to write the spinorial field strength as [125]

$$F_{ABA'B'} = \frac{1}{2}\left(F_{ABA'B'} - F_{ABB'A'}\right) + \frac{1}{2}\left(F_{ABB'A'} - F_{BAB'A'}\right), \tag{5.34}$$

where the first (second) contribution is antisymmetric in the primed (unprimed) indices. Using the earlier fact that a quantity antisymmetric in a given pair of indices must be proportional to an appropriate Levi-Civita symbol, we thus find that the spinorial translation of the field strength must have the general form

$$F_{ABA'B'} = \phi_{AB}\epsilon_{A'B'} + \bar{\phi}_{A'B'}\epsilon_{AB}. \tag{5.35}$$

Given that taking the complex conjugate of a spinor interchanges primed and unprimed indices, one finds that if the field strength is real, then $\bar{\phi}_{A'B'}$ must indeed be the complex conjugate of ϕ_{AB}, as suggested by our notation. There is also a further interpretation one may give to the spinors ϕ_{AB} and $\bar{\phi}_{A'B'}$. For any field strength tensor, one may define the *dual field strength*

$$\tilde{F}_{\mu\nu} = \frac{1}{2}\epsilon_{\mu\nu\alpha\beta}F^{\alpha\beta}, \qquad (5.36)$$

where $\epsilon_{\mu\nu\alpha\beta}$ is the four-dimensional Levi-Civita symbol. As may be shown by direct calculation, this has the effect of interchanging the electric and magnetic fields (up to a minus sign), and we may then split a given field strength into two parts, each of which has simple properties when constructing the dual tensor:

$$F_{\mu\nu} = F^+_{\mu\nu} + F^-_{\mu\nu}, \quad \tilde{F}^\pm_{\mu\nu} = \pm i F^\pm_{\mu\nu}. \qquad (5.37)$$

The quantity $F^+_{\mu\nu}$ ($F^-_{\mu\nu}$) is called the *self-dual* (*anti-self-dual*) part of $F_{\mu\nu}$ and will necessarily be complex if the field strength itself is real. This itself has a very physical interpretation: for plane wave solutions, the (anti-)self-dual parts of the field can be shown to be none other than the two independent circular polarisation states of light, where a complex polarisation vector is needed to keep track of the phase difference between two constant (linear) polarisation states. Returning to our spinor formula in Eq. (5.35), it may be shown (see e.g. Ref. [125]) that the two terms correspond, respectively, to the anti-self-dual and self-dual parts of the field. Finally, we note that the decomposition of Eq. (5.35) makes sense from the point of view of counting degrees of freedom. The field strength tensor $F_{\mu\nu}$ has six real degrees of freedom, corresponding to the three physical components each of the electric and magnetic fields. The spinors ϕ_{AB} and $\bar{\phi}_{A'B'}$ have six complex degrees of freedom between them, which is reduced to six real degrees of freedom upon demanding that $\bar{\phi}_{A'B'}$ is the complex conjugate of ϕ_{AB}. From now on, we refer to ϕ_{AB} ($\bar{\phi}_{A'B'}$) as the *electromagnetic (dual) spinor*. We can translate the vacuum field equation of Eq. (2.30) (i.e. with $j_\nu = 0$) by using Eq. (5.26), and we ultimately find

$$\nabla^{AA'}\phi_{AB} = 0, \quad \nabla^{AA'}\bar{\phi}_{A'B'} = 0, \qquad (5.38)$$

where we have introduced the spinorial translation of the partial derivative:

$$\nabla_{AA'} = \sigma^\mu_{AA'}\partial_\mu. \qquad (5.39)$$

For a real electromagnetic field, the second equation in Eq. (5.38) can simply be obtained as the complex conjugate of the other.

We can carry out a similar analysis for gravity, although the details will necessarily be more complicated given that the analogue of the electromagnetic field strength tensor is the Riemann tensor $R_{\mu\nu\alpha\beta}$, which has four spacetime indices. Its spinorial translation thus has eight spinor indices. However, one may use known symmetries of the Riemann tensor under interchange of indices, and the earlier property of reduction to symmetric spinors, to show that the general form of the Riemann tensor in the spinorial language is (see e.g. Refs. [125, 126])

$$
\begin{aligned}
R_{ABCDA'B'C'D'} = {} & \Psi_{ABCD}\epsilon_{A'B'}\epsilon_{C'D'} + \bar{\Psi}_{A'B'C'D'}\epsilon_{AB}\epsilon_{CD} \\
& + \epsilon_{A'B'}\epsilon_{CD}\Phi_{ABC'D'} + \epsilon_{AB}\epsilon_{C'D'}\bar{\Phi}_{A'B'CD} \\
& - 2\Lambda \left[\epsilon_{A'B'}\epsilon_{C'D'}\epsilon_{A(C}\epsilon_{D)B} \right. \\
& \left. + \epsilon_{AB}\epsilon_{CD}\epsilon_{A'(C'}\epsilon_{D')B')} \right].
\end{aligned}
\tag{5.40}
$$

As earlier, the barred quantities are related to their unbarred counterparts by complex conjugation, if the Riemann tensor is real. Also, Λ is proportional to the cosmological constant, such that the terms in the third line will vanish if this is zero. One may also show that the terms in the second line vanish if one considers vacuum solutions of the Einstein equations (i.e. those for which the Einstein tensor comprising the left-hand side of Eq. (2.105) is zero). This leaves only the terms in the first line of Eq. (5.40). In the tensor language for vacuum spacetimes, it is known that the Riemann tensor reduces to a quantity called the *Weyl tensor*, which is usually written as $C_{\mu\nu\alpha\beta}$. We may thus write Eq. (5.40) in the vacuum case as

$$
C_{ABCDA'B'C'D'} = \Psi_{ABCD}\epsilon_{A'B'}\epsilon_{C'D'} + \bar{\Psi}_{A'B'C'D'}\epsilon_{AB}\epsilon_{CD}, \tag{5.41}
$$

where the quantity Ψ_{ABCD} ($\bar{\Psi}_{A'B'C'D'}$) is known as the *Weyl (dual) spinor*. A non-zero Weyl spinor represents a deviation of a vacuum spacetime from Minkowski space and thus the presence of gravity! Note that, despite the differing number of indices, Eq. (5.41) nevertheless bears a resemblance to the field strength of Eq. (5.35). That is, there are two terms related by complex conjugation, and one may then ask if they can be furnished with a similar physical interpretation. Indeed they can: if we define the *dual Riemann tensor* by (cf. Eq. (5.36))

$$\tilde{R}_{\mu\nu\alpha\beta} = \frac{1}{2}\epsilon_{\alpha\beta}{}^{\rho\lambda}R_{\mu\nu\rho\lambda}, \tag{5.42}$$

we can then split the Riemann tensor into (anti-)self-dual parts (cf. Eq. (5.37))

$$R_{\mu\nu\alpha\beta} = R_{\mu\nu\alpha\beta}^{+} + R_{\mu\nu\alpha\beta}^{-}, \quad \tilde{R}_{\mu\nu\alpha\beta}^{\pm} = \pm i R_{\mu\nu\alpha\beta}, \tag{5.43}$$

which straightforwardly applies also to the Weyl tensor. One may then show that Φ_{ABCD} ($\bar{\Phi}_{A'B'C'D'}$) corresponds with the anti-self-dual (self-dual) part of the Weyl tensor, and we can then interpret the two terms in Eq. (5.41) as representing the two independent polarisation states of a propagating gravitational field. One may also translate the vacuum Einstein equation into a constraint on the Weyl (dual) spinor, and the result is [125, 126][3]

$$\nabla^{AA'}\Psi_{ABCD} = 0, \quad \nabla^{AA'}\bar{\Psi}_{A'B'C'D'} = 0. \tag{5.44}$$

Again, this bears an uncanny resemblance to its electromagnetic analogue, Eq. (5.38). Both are special cases of the so-called *massless free field equation*

$$\nabla^{AA'}\psi_{AB...C}, \quad \nabla^{AA'}\bar{\psi}_{A'B'...C'} = 0, \tag{5.45}$$

where there are $2n$ indices in general for a field of spin n.

5.3 The Petrov Classification

In the previous section, we have seen that vacuum solutions of electromagnetism and gravity can be expressed in terms of spinors ϕ_{AB} and Ψ_{ABCD} (and their complex conjugates), respectively, where these are symmetric in their indices. However, we also saw in Section 5.1.2

---

[3]Strictly speaking, $\nabla^{AA'}$ in Eq. (5.44) should be the translation of the covariant derivative associated with a given spacetime, rather than the partial derivative. However, we only need to focus on the simpler case in what follows.

that any symmetric spinor can be decomposed into principal spinors so that we may in general write

$$\phi_{AB} = \alpha_{(A}\beta_{B)}, \quad \Psi_{ABCD} = \eta_{(A}\lambda_B\gamma_C\delta_{D)}. \quad (5.46)$$

This allows us to classify solutions into qualitatively different types. For the electromagnetic spinor, for example, we may draw a distinction between solutions for which the two principal spinors are proportional ($\alpha \propto \beta$) and those where they are not ($\alpha \not\propto \beta$). Solutions of the former type are called *null*, and those of the latter type *non-null*. Although there are infinitely many solutions of each type, this classification is nevertheless useful.

For gravity, there are many more possibilities than the earlier two choices, due to the fact that the Weyl spinor has four spinor indices rather than two. These are listed in Table 5.1, and this classification of solutions of general relativity based on the structure of the Weyl spinor is known as the *Petrov classification*. In the table, we indicate the different possible patterns of degeneracy of the principal spinors where e.g. $\{2, 2\}$ denotes that there are two distinct pairs of principal spinors, where the spinors in each pair are mutually proportional. Corresponding to each degeneracy pattern is a label known as the *Petrov type*, and these labels are still in common use.

Although the same classification can be discussed in the tensor language (see e.g. [127, 128] for textbook treatments), it is much more cumbersome. The spinor language makes this structure manifest, which is a direct consequence of the simplification properties discussed in Section 5.1.2, each of which relied upon the fact that spinor

Table 5.1. Different types of Weyl spinors classified by (i) the pattern of degenerate principal null directions and (ii) the equivalent Petrov type.

Weyl Type	Petrov Label
$\{1, 1, 1, 1\}$	I
$\{2, 1, 1\}$	II
$\{3, 1\}$	III
$\{4\}$	N
$\{2, 2\}$	D
$\{-\}$	O

indices can only take two distinct values. Here, we have explicitly discussed vacuum solutions of electromagnetism and/or gravity. However, the classification schemes described earlier can also be applied to non-vacuum solutions. In the gravity case, one still classifies solutions according to principal spinors of the Weyl spinor, even in the non-vacuum case.

5.4 The Weyl Double Copy

We are now ready to state the Weyl double copy. This was originally presented in Ref. [124], and one may also find earlier results in the general relativity literature that partially anticipate the result [129–131], which is as follows. Given an electromagnetic field strength spinor ϕ_{AB}, one can construct a gravitational Weyl spinor according to the rule

$$\Psi_{ABCD}(x) = \frac{1}{S(x)} \phi_{(AB}(x)\phi_{CD)}(x), \tag{5.47}$$

where S is a scalar field satisfying

$$\partial^\mu \partial_\mu S(x) = 0. \tag{5.48}$$

In Eq. (5.47), we have emphasised that all (spinor) fields are functions of spacetime position in general and that the Weyl double copy applies point-by-point. An analogous relationship applies for dual spinor quantities, where the scalar field will be replaced by its complex conjugate $\bar{S}(x)$:

$$\bar{\Psi}_{A'B'C'D'}(x) = \frac{1}{\bar{S}(x)} \bar{\phi}_{(A'B'}(x)\bar{\phi}_{C'D')}(x). \tag{5.49}$$

To see that these equations are plausible, let us first note that the Weyl spinor must be symmetric in its indices and that this is indeed satisfied in Eqs. (5.47) and (5.49) due to the explicit symmetrisation over indices on the right-hand sides. Furthermore, the relationship is reminiscent of the Kerr–Schild double copy and the BCJ double copy for scattering amplitudes, in that two copies of quantities from a gauge theory are combined to make a gravity result. An apparent difference with respect to the Kerr–Schild case is that we now have a scalar field in the denominator of the result for the gravity solution. This is due, however, to the way in which we presented the Kerr–Schild double copy in the previous chapter. The Kerr–Schild double copy of an electromagnetic field $A_\mu = \phi k_\mu$ has the form

$$h_{\mu\nu} = \phi k_\mu k_\nu = \frac{1}{\phi} A_\mu A_\nu,$$

such that writing the graviton in terms of electromagnetic fields indeed yields a scalar field in the denominator that compensates for the fact that the scalar has been counted twice in the numerator. Viewed in this way, the Weyl double copy seems to be a close counterpart of its Kerr–Schild cousin, where the scalar field $S(x)$ in the former might be related to the field $\phi(x)$ in the latter. We examine this relationship more precisely in the following.

The Weyl double copy was argued in Ref. [124] to apply to arbitrary vacuum Petrov type D solutions, for which a general family is known to exist [132]. A crucial feature of these solutions is that they are known to precisely linearise the Einstein equations, such that Eq. (5.44) constitutes an exact solution for a given spacetime. Likewise, the corresponding spinor ϕ_{AB} will constitute an exact solution of electromagnetism, given that this theory is linear anyway. This is analogous to how, in the Kerr–Schild double copy, the solutions that could be double-copied all satisfied linear versions of their respective field equations. Reference [124] gave several explicit examples of Weyl double copies, including a physical interpretation of the single copy i.e. the electromagnetic spinor that enters Eq. (5.47). For our own illustrative purposes, the relatively simple case of the Kerr black hole suffices. Rather than the Kerr–Schild coordinates encountered in the previous chapter, Ref. [124] presented the Kerr metric in more conventional *Boyer–Lindquist coordinates* (t, r, θ, ψ), which are, respectively, a time coordinate, a radial coordinate, and polar and azimuthal angle coordinates. The Kerr line element may then be written as

$$ds^2 = -dt^2 + \Sigma^2 \left(\frac{dr^2}{\Delta} + \frac{dX^2}{1 - X^2} \right) + (r^2 + a^2)(1 - X^2)d\psi^2$$

$$+ \frac{2Mr}{\Sigma^2} \left(dt + a(1 - X^2)d\psi \right)^2, \tag{5.50}$$

with

$$\Sigma^2 = r^2 + a^2 X^2, \quad \Delta = r^2 + a^2 - 2Mr, \quad X = \cos\theta. \tag{5.51}$$

As before, a is related to the angular momentum of the black hole and M is its mass. In terms of these coordinates, we may define the electromagnetic spinor

$$\Phi_{AB} = \frac{Q}{2(r + iaX)^2} \alpha_{(A}\beta_{B)}, \quad \alpha_A = (-1,1), \quad \beta_A = (1,1), \tag{5.52}$$

as well as the Weyl spinor

$$C_{ABCD} = -\frac{3M}{2(r + iaX)^3} \alpha_{(A}\alpha_B\beta_C\beta_{D)}. \tag{5.53}$$

These expressions can be seen to obey the Weyl double-copy formula of Eq. (5.47), provided the scalar function is given by

$$S = -\frac{Q^2}{6M} \frac{1}{r + iaX}, \tag{5.54}$$

which indeed satisfies the wave equation of Eq. (5.48) in Minkowski spacetime.[4] Note that we have followed Ref. [124] in defining the scalar function precisely so as to replace factors of electromagnetic charge Q with the mass M in gravity. One could also take a different approach and explicitly replace charge factors as one does in the Kerr–Schild double copy. Each solution is anyway defined only up to a constant numerical factor, due to being a solution of the relevant massless free field equation. Similar expressions to the above are obtained for the dual Weyl spinor. One finds that the principal spinors are the same, but the scalar function will now be replaced by its complex conjugate:

$$\bar{S} = -\frac{Q^2}{6M} \frac{1}{r - iaX}. \tag{5.55}$$

---

[4]In showing this, it is important to evaluate the operator ∂^2 in the Boyer–Lindquist coordinate system adopted in Eq. (5.54).

A corollary of this result is that we obtain the Schwarzschild solution as $a \to 0$, such that the scalar function becomes simply

$$S \propto \frac{1}{r}.$$

Since its inception, the Weyl double copy has also been applied to type N solutions relevant for gravitational waves [133], as well as novel solutions of potential relevance to optics and condensed matter [134] that we discuss in the following. Further work has looked at relationships between the gauge theory field strength and Weyl tensor in the traditional (non-spinor) language [135, 136], including generalisations to higher dimensions. An alternative approach to exploring the double copy using spinorial ideas is described in Ref. [137].

Earlier, we have considered the Weyl double copy for the case in which both electromagnetic spinors entering Eq. (5.47) are the same. As pointed out in Ref. [124], however, one can also consider the more general formula

$$\Psi_{ABCD} = \frac{\Phi^{(1)}_{(AB} \Phi^{(2)}_{CD)}}{S}, \tag{5.56}$$

where there are now two different electromagnetic spinors in the numerator. This is usually called the *mixed Weyl double copy* and clearly works for the type D solutions considered earlier. In the Kerr case, for example, we could have chosen

$$\Phi^{(1)}_{AB} = \frac{Q}{2(r + iaX)^2} \alpha_A \alpha_B, \quad \Phi^{(2)}_{AB} = \frac{Q}{2(r + iaX)^2} \beta_A \beta_B, \tag{5.57}$$

such that Eq. (5.56) gives the same Weyl spinor as Eq. (5.47). This also then raises the question of whether other Petrov types are possible, at least at linear level in the field equations. A more underlying concern is where the Weyl double copy comes from in the first place and whether it can be formally related to the original BCJ double copy for scattering amplitudes, or to the Kerr–Schild double copy. A recent series of papers [138–141] (see also [142]) has explored this systematically, using a novel set of mathematical methods that we describe in the following section.[5]

---

[5]See also Refs. [143, 144] for work relating the Weyl double copy to scattering amplitudes.

5.5   The Twistor Double Copy

Twistor theory is a decades old set of mathematical ideas combining algebraic geometry and complex analysis [145–147] (see Refs. [125, 148, 149] for pedagogical reviews and Ref. [150] for a more modern viewpoint). The basic idea of twistor theory is that objects in spacetime can be mapped to an abstract *twistor space*, and vice versa. This provides an alternative viewpoint on spacetime physics, and the relevance for this chapter is that twistorial insights can be used to resolve many puzzles regarding the Weyl double copy.

There are various different ways to introduce twistors. Here we follow an approach inspired by Ref. [151], which is perhaps more accessible for beginners, and for the brief review that is necessitated here. We start by asserting that various elements of twistor theory are more natural to appreciate if we work in a complexified version of Minkowski space, which we will denote by $\mathbb{M}_C$. That is, we work with the line element of Eq. (4.2) but take all coordinates appearing there to be complex, rather than real. We can then work towards describing what a twistor is by focusing on certain families of planes in $\mathbb{M}_C$.

5.5.1   *Null planes*

Recall in ordinary vector analysis in three dimensions that the equation of a plane can be written as

$$n \cdot x = a, \qquad (5.58)$$

where n is a normal vector to the plane and a a constant.[6] The normal vector n can be obtained by finding two vectors e_1 and e_2 lying in the plane and forming the combination

$$n = e_1 \times e_2. \qquad (5.59)$$

That is, the cross-product of e_1 and e_2 must be perpendicular to both vectors and thus perpendicular to the plane of interest. These considerations do not generalise immediately to higher numbers of dimensions, given that the cross-product of two vectors can no longer be

---

[6]If n is a unit vector, then the constant a appearing on the right-hand side of Eq. (5.58) turns out to be the perpendicular distance of the plane from the origin.

defined. Instead, one may consider an antisymmetric matrix defined in terms of the two vectors spanning the plane:

$$\omega_{ij} = e_{1i}e_{2j} - e_{1j}e_{2i}, \qquad (5.60)$$

where indices on the right-hand side denote 3-vector components. The components of the normal vector – itself a quantity unique to $D = 3$ space dimensions, given that the "normal" direction becomes ambiguous for $D > 3$ – can be defined in terms of ω_{ij} as follows:

$$n_i = \frac{1}{2}\epsilon_{ijk}\omega_{jk}, \qquad (5.61)$$

where ϵ_{ijk} is the three-dimensional Levi-Civita symbol. However, we may regard ω_{ij} as the more fundamental quantity, given that its definition clearly generalises to any number of dimensions and also to Minkowski spacetime. It is called the *tangent bivector* to the plane, and the equivalent of this for a plane in $\mathbb{M}_C$ is

$$\omega_{\mu\nu} = e_{1\mu}e_{2\nu} - e_{1\nu}e_{2\mu}, \qquad (5.62)$$

where now $e_{i\mu}$ is a 4-vector (with complex components) lying in the plane of interest. The tangent bivector is by definition antisymmetric in its spacetime indices. However, we saw in Section 5.2 that any antisymmetric 2-index tensor could be decomposed into a sum of (anti-)self-dual parts, defined in the present case by replacing $F_{\mu\nu}$ with $\omega_{\mu\nu}$ in Eq. (5.37). Planes are called (anti-)self-dual if their tangent bivector has the corresponding property, and a general plane in $\mathbb{M}_C$ will be neither self-dual nor anti-self-dual. However, the condition of being (anti-)self-dual imposes additional restrictions. To see this, note that the (anti-)self dual condition on $\omega_{\mu\nu}$ can be written as

$$\frac{1}{2}\epsilon_{\mu\nu\alpha\beta}\left[e_1^\alpha e_2^\beta - e_2^\beta e_1^\alpha\right] = \pm i \left[e_{1\mu}e_{2\nu} - e_{2\mu}e_{1\nu}\right].$$

If we contract the left-hand side with either e_1^μ or e_2^μ, it vanishes due to antisymmetry of the Levi-Civita symbol. This leads to the following two conditions:

$$e_1^2 e_2^\mu - (e_1 \cdot e_2)e_1^\mu = 0, \quad (e_1 \cdot e_2)e_2^\nu - e_2^2 e_1^\nu = 0.$$

We can choose the two vectors lying in the plane to be orthogonal, which then implies

$$e_1 \cdot e_2 = e_1^2 = e_2^2 = 0, \tag{5.63}$$

i.e. that e_1^μ and e_2^ν must be null vectors. A general vector lying in the plane will be given by

$$e^\mu = k_1 e_1^\mu + k_2 e_2^\mu,$$

for some constants $\{k_i\}$, and this will also then be null, from Eq. (5.63). We are thus led to the conclusion that (anti-)self-dual planes are *null planes*, meaning that every tangent vector to the plane is null. In the literature on twistor theory, self-dual and anti-self-dual planes are known as α-planes and β-planes, respectively, and we use this terminology in what follows.

5.5.2 *Twistors and dual twistors*

In terms of the earlier notation, the equation for a null plane can be written as

$$x^\mu = x_0^\mu + k_1 e_1^\mu + k_2 e_2^\mu, \tag{5.64}$$

where x_0^μ is any point lying in the plane, and the remaining terms on the right-hand side form a null vector. We can find the spinorial translation of Eq. (5.64) by contracting with the Infeld–van der Waerden symbols, and the result is

$$x^{AA'} = x_0^{AA'} + \lambda^A \pi^{A'}, \tag{5.65}$$

where $x_{0\,AA'}$ is the spinorial translation of x_0^μ, and we have used the fact that the spinor matrix corresponding to any null vector can be factorised into an outer product of two spinors (n.b. it has zero determinant). Unlike in Eq. (5.4), however, the two spinors (here labelled λ_A and $\pi_{A'}$) need not be related by complex conjugation, given that we are in complexified Minkowski spacetime. Also, whether or not we are talking about spinors or dual spinors is simply labelled by whether or not there is a prime on the spinor index.

A given α-plane can be generated from Eq. (5.65) by taking $\pi_{A'}$ to be fixed and varying λ_A. To see this, note that upon making this choice, the spinorial translation of our two basis vectors in the plane can be written as

$$e_1^\mu \to \lambda_{1A}\pi_{A'}, \quad e_2^\mu \to \lambda_{2A}\pi_{A'}. \tag{5.66}$$

We then have

$$e_1^\mu e_2^\nu - e_2^\mu e_1^\nu \to (\lambda_{1A}\lambda_{2B} - \lambda_{2A}\lambda_{1B})\,\pi_{A'}\pi_{B'}. \tag{5.67}$$

Using the fact that the antisymmetric combination in the λ spinors must be proportional to the Levi-Civita symbol (cf. Section 5.1.2), we find that the spinorial translation of the tangent bivector is

$$\omega_{ABA'B'} \propto \epsilon_{AB}\pi_{A'}\pi_{B'}. \tag{5.68}$$

Comparison with Eq. (5.35) and the discussion thereunder shows that our tangent bivector is self-dual, and we thus have an α-plane as required. Different choices of $\pi_{A'}$ correspond to different α-planes, and we can write Eq. (5.65) in a more convenient form that eliminates λ_A by contracting with $\pi_{A'}$:

$$\omega^A = ix^{AA'}\pi_{A'}, \tag{5.69}$$

where we have defined[7]

$$\omega^A = ix_0^{AA'}\pi_{A'}, \tag{5.70}$$

and the factor of i is conventional. To fully specify a given α-plane, we need to give the values of the following: (i) $\pi_{A'}$, which controls the orientation of the plane (i.e. it enters all possible tangent vectors); (ii) ω^A, which is related to $x_0^{AA'}$, and thus to where the plane is located relative to the origin of $\mathbb{M}_C$. We can collect these two quantities into a single 4-component object:

$$Z^a = \left(\omega^A, \pi_{A'}\right) \equiv (Z^0, Z^1, Z^2, Z^3), \tag{5.71}$$

which is called a *twistor*. We have here used a lower-case Latin index to denote which component we are talking about, to avoid confusion

---

[7]Note that $x^{AA'}$ in Eq. (5.69) contains the full dependence of Eq. (5.65), but where the second term vanishes to leave only the dependence on the reference point $x_0^{AA'}$.

with the Greek letters used for spacetime indices, and the capital Latin letters used for spinor components![8] Strictly speaking, a twistor is defined such that ω^A and $\pi_{A'}$ are independent from each other, and *twistor space* $\mathbb{T}$ is defined to be the set of all such objects. We only need to consider twistors obeying Eq. (5.69), which is known as the *incidence relation*. Looking at this, we see that the two "halves" of the twistor appear on the left- and right-hand sides, respectively. Thus, scaling all components of the twistor by

$$Z^a \to \lambda Z^a, \quad \lambda \in \mathbb{C}$$

leaves the incidence relation invariant. A fancy way of saying this is that twistors obeying the incidence relation correspond to points in *projective twistor space*, which is usually denoted by $\mathbb{PT}$. The word "projective" means that the components of the twistor are defined only up to an arbitrary complex scale factor. All of this is highly abstract, but a way to "visualise" $\mathbb{PT}$, as introduced earlier, is that it corresponds to the space of all possible α-planes in spacetime. That is, a single point in $\mathbb{PT}$ corresponds to a given α-plane in $\mathbb{M}_C$. As we move around $\mathbb{PT}$, we vary the quantities ω^A and $\pi_{A'}$, which corresponds to picking out different α-planes in $\mathbb{M}_C$. Note that it is the incidence relation of Eq. (5.69) that creates an explicit map between spacetime (as represented by $x^{AA'}$) and projective twistor space. Furthermore, the fact that points in $\mathbb{PT}$ are mapped to extended objects in spacetime (α-planes) means that the map between the two spaces is non-local.

We can also ask what points in spacetime ($\mathbb{M}_C$) correspond to in projective twistor space ($\mathbb{PT}$). To see this, note that a general twistor starts out with four complex degrees of freedom. Fixing a spacetime point $x^{AA'}$ in the incidence relation of Eq. (5.69) then imposes two complex constraints, telling us that ω^A is known once $\pi_{A'}$ is chosen. Finally, there is an overall scaling factor that we are allowed to extract from $\pi_{A'}$, and thus we can consider $\pi_{A'} = (1, \xi)$ or $\pi_{A'} = (\eta, 1)$, where $\xi, \eta \in \mathbb{C}$ are allowed to be zero. We need both of these parametrisations to cover all possible choices of $\pi_{A'}$, and thus our point in spacetime maps to a surface in $\mathbb{PT}$ parametrised

---

[8]More common in the twistor literature is that lower-case Latin and Greek letters are used for spacetime and twistor indices, respectively.

by two coordinate patches, each of which can be mapped to the complex plane. This is precisely the Riemann sphere we discussed in Section 5.1.1, and thus a spacetime point x^μ maps to a Riemann sphere X in projective twistor space. It is also sometimes referred to as a *(complex) line* in twistor space given that there is a single complex degree of freedom. In particular, literature on twistor methods for scattering amplitudes often draws lines to represent spacetime points that have been mapped to $\mathbb{PT}$, and this visual representation is useful when thinking about how the surfaces corresponding to *different* spacetime points can intersect.

Earlier, we have focused only on α-planes. As may already be clear from Eq. (5.65), we could also have made the choice to fix λ^A and vary $\pi_{A'}$. Carrying through similar arguments leads to an anti-self-dual bivector

$$\omega_{ABA'B'} \propto \lambda_A \lambda_B \epsilon_{A'B'}$$

and thus to a β-plane. Contracting Eq. (5.65) with λ_A, we can write the equation of such a plane as

$$\mu^{A'} = -ix^{AA'}\lambda_A, \tag{5.72}$$

where we have defined

$$\mu^{A'} = -ix_0^{AA'}\lambda_A. \tag{5.73}$$

We can now group the quantities $\mu^{A'}$ and λ_A into a single 4-component object called a *dual twistor*:

$$W_a = \left(\lambda_A, \mu^{A'}\right), \tag{5.74}$$

where, following convention, we place the unprimed spinor first, followed by the primed one (as in Eq. (5.71)). As in the twistor case, Eq. (5.72) is known as the incidence relation, and we have included a factor of $-i$ on the right-hand side, which turns out to be convenient for the following reason. Upon returning from complexified Minkowski space to conventional (real) spacetime, the two incidence relations of Eqs. (5.69) and (5.72) turn out to be related by complex conjugation, analogous to how spinors can be turned into dual spinors. Dual twistors obeying Eq. (5.72) are said to be points in *projective dual twistor space*, denoted by $\mathbb{PT}^*$. Points in the latter constitute β-planes in $\mathbb{M}_C$. Similar to the twistor case, a point x^μ in $\mathbb{M}_C$ is associated with a Riemann sphere X in $\mathbb{PT}^*$.

Earlier, we have represented (dual) twistors with upstairs (downstairs) indices, respectively. There is no explicit need to do this, given that there is no metric in twistor space with which to raise and/or lower indices. However, this index placement is useful in that it reminds us what kind of twistor (dual or otherwise) we are talking about. We may also define the following *inner product* for (dual) twistors:

$$Z^a W_a = \omega^A \lambda_A + \mu^{A'} \pi_{A'} \in \mathbb{C}. \tag{5.75}$$

This product turns out to be invariant under the conformal transformations of spacetime that we introduced in the previous chapter. Indeed, an alternative – and even more mathematical – way of introducing twistors is that they are the natural objects that transform linearly under conformal transformations, analogous to how 4-vectors and tensors are the objects that transform linearly under Lorentz transformations (n.b. they transform non-linearly under conformal transformations in general). Twistor variables have become increasingly prevalent in scattering amplitude research over the past two decades, starting with the seminal work of Ref. [152]. Much of the reason for this is that many theories considered in contemporary amplitudes research have conformal symmetry built in, thus making twistor variables the natural quantities to use.

5.5.3 *The Penrose transform*

We have seen that spacetime points can be mapped non-locally to objects in twistor space, and vice versa. In order to address the double copy, we need to see what spacetime (spinor) fields look like in twistor space, and there is a central result of twistor theory that provides the correspondence we need. Called the *Penrose transform*, it takes the form of a certain integral in (projective) twistor space that turns out to yield a field in spacetime. Let us consider a function $f(Z^a)$ that depends on twistor coordinates but no non-constant dual twistors. In order to build in a spacetime dependence, let us denote by $\rho_x[f(Z^a)]$ the restriction to the Riemann sphere X in twistor space corresponding to the spacetime point x^μ. In practical terms, this means that the twistor Z^a is understood to be obeying the incidence relation of Eq. (5.69). We can then consider the following integral:

$$\phi_{A'B'...C'}(x) = \frac{1}{2\pi i} \oint_\Gamma \pi_{E'} d\pi^{E'} \pi_{A'} \pi_{B'} \cdots \pi_{C'} \rho_x[f(Z^a)], \qquad (5.76)$$

where Γ is a closed contour on the Riemann sphere X and $\pi_{E'}$ is the spinor that enters the twistor Z^a, as in Eq. (5.71). At first glance, it is not at all obvious what this integral is meant to correspond to. However, the left-hand side is clearly some spinor quantity that depends on spacetime position, given that the right-hand side is defined in terms of (integrals of) spinors, and also includes a manifest x-dependence. Furthermore, the spinor on the left-hand side must be symmetric in all its indices, as can be seen by inspecting the right-hand side. Thus, the quantity $\phi_{A'B'...C'}$ has the right properties to potentially correspond to a spacetime spinor field. We may go further than this in showing that it in fact satisfies the relevant massless free field equation of Eq. (5.45). First, note that we have

$$\nabla_{DD'}\rho_x[f(Z^a)] = \frac{\partial}{\partial x^{DD'}} f(ix^{AA'}\pi_{A'}, \pi_{A'})$$

$$= i\pi_{D'}\rho_x\left[\frac{\partial f(Z^a)}{\partial\omega^D}\right],$$

where we have used the incidence relation of Eq. (5.69). Equation (5.76) then implies

$$\nabla_{DD'}\phi_{A'B'...C'} = \frac{1}{2\pi} \oint_\Gamma \pi_{E'} d\pi^{E'} \pi_{A'} \pi_{B'} \cdots \pi_{C'} \pi_{D'} \rho_x\left[\frac{\partial f(Z^a)}{\partial\omega^D}\right].$$

$$(5.77)$$

To form the massless free field equation from Eq. (5.45), we must contract the D' and A' indices on the left-hand side using $\epsilon^{A'D'}$. However, the right-hand side is symmetric in A' and D' and thus vanishes upon contraction with the antisymmetric $\epsilon^{A'D'}$. Hence, Eq. (5.45) follows as required. Crucial in this analysis is that there was no dependence on non-constant dual twistor coordinates, which would have led to both incidence relations of Eqs. (5.69) and (5.72) contributing and would have spoilt the final conclusion. We thus see that twistor functions $f(Z^a)$ correspond to self-dual (primed) spinor fields in spacetime. Likewise, we may write a corresponding formula for anti-self-dual fields, based on dual twistor coordinates[9]:

$$\phi_{AB...C}(x) = \frac{1}{2\pi i} \oint_\Gamma \lambda_E d\lambda^E \lambda_A \lambda_B \ldots \lambda_C \rho_x[f(W_a)]. \tag{5.78}$$

We thus see that functions of (dual) twistor coordinates are related to (anti-)self-dual spacetime fields, respectively. Given that the massless free field equation is linear, we can superpose such solutions in spacetime to obtain more general field solutions.

Earlier, we have loosely referred to the quantities $f(Z^a)$ as *functions*, but in fact they are a more complicated type of mathematical object. The basic reason for this is that we have considerable freedom to redefine $f(Z^a)$ in Eq. (5.76). We depict the Riemann sphere X in Figure 5.2, and the contour Γ which, by reparametrising the sphere, we can choose to coincide with the equator. Let us then parametrise

$$\pi_{A'} = (1, \xi), \quad \xi \in \mathbb{C} \tag{5.79}$$

corresponding to a stereographic projection from the north pole, which maps the sphere X to the complex plane of ξ, where the south and north poles of X are at $\xi = 0$ and $\xi = \infty$. Figure 5.2(a) then translates as Figure 5.2(b), where the northern and southern hemispheres now correspond to outside and inside the contour Γ, respectively. The Penrose transform now becomes a conventional contour integral in the parameter ξ, which may be evaluated using Cauchy's formula. To obtain a non-zero answer, we must have singularities of the twistor function $f(Z^a)$ both inside and outside Γ. Otherwise, we can simply choose to apply Cauchy's formula to the region N or S that has no singularities, obtaining a zero answer. In terms of X itself, we find that $f(Z^a)$ must have singularities in both hemispheres. A further consequence of this discussion is that we can redefine $f(Z^a)$

---

[9]An alternative Penrose transform exists for unprimed spinor fields, in which twistor functions $f(Z^a)$ are acted upon with derivative operators with respect to ω^A. This has the advantage that everything can be expressed in terms of twistor space rather than having to invoke dual twistors, but we will not need this alternative Penrose transform here. See e.g. Ref. [148] for a comprehensive discussion and Ref. [139] within a double-copy context.

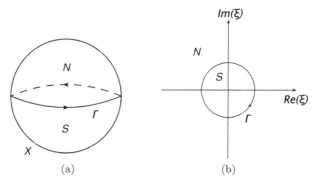

Fig. 5.2. (a) The Penrose transform contour Γ separates the Riemann sphere X corresponding to a spacetime point x^μ into its northern and southern hemispheres N and S; (b) stereographic projection of part (a) onto the complex plane of a variable ξ.

according to the following transformation:

$$f(Z^a) \to f(Z^a) + f_N(Z^a) + f_S(Z^a), \qquad (5.80)$$

where $f_N(Z^a)$ $(f_S(Z^a))$ has singularities only in N (S), respectively. The additional terms, when substituted into Eq. (5.76), will give a zero answer after carrying out the integral. We are to regard all such choices of $f(Z^a)$ as equivalent from the point of view of the Penrose transform: after all, they all give rise to the same spacetime field. Mathematically, the set of functions $\{f(Z^a)\}$ that are equivalent under Eq. (5.80) is called a *cohomology class*, and a given choice $f(Z^a)$ is a *representative* of such a class. One may formally prove [153] that the set of all spacetime fields is in one-to-one correspondence with the appropriate set of cohomology classes in $\mathbb{PT}$. We will not elaborate on this here but point to Ref. [140] for a pedagogical discussion in the recent double-copy literature. In what follows, we incur the potential wrath of mathematicians by continuing to use the vulgar slang word "function" to mean "representative of cohomology class", not least to save ink.

We have so far talked as if the twistor function $f(Z^a)$ that enters Eq. (5.76) is completely arbitrary. In fact, there is an important constraint on it, and its equivalent functions under Eq. (5.80), that arises from the fact that the Penrose transform is in *projective* twistor space. As we saw earlier, we are meant to be able to rescale the twistor Z^a by an arbitrary constant, which in turn rescales $\pi_{A'}$, without this having any consequences. For this to make sense in Eq. (5.76), it must be the case that $f(Z^a)$ obeys

$$f(\lambda Z^a) = \lambda^{-2n-2} f(Z^a), \quad \lambda \in \mathbb{C}, \tag{5.81}$$

where $2n$ is the number of indices on the left-hand side, and n the spin of the corresponding field. Provided Eq. (5.81) is true, the total integrand in Eq. (5.76), including the measure, is invariant under rescalings as required. We say that $f(Z^a)$ is *homogeneous* and that it has *homogeneity* $-2n - 2$. Earlier, we have not explicitly discussed the case of scalar fields. From Eq. (5.76), we expect this to be given by integrals of the form

$$\phi(x) = \frac{1}{2\pi i} \oint_\Gamma \pi_{E'} d\pi^{E'} \rho_x[f(Z^\alpha)], \tag{5.82}$$

and one may show that this indeed satisfies the massless Klein–Gordon equation [148].

5.5.4   *The Weyl double copy from twistor space*

We now have all the ingredients we need to derive the Weyl double copy from twistor space. Our starting point is the observation made above that spacetime fields with different spins can be identified with twistor functions with different homogeneities. Consider taking two homogeneity -4 functions $f_{\text{EM}}^{(1)}(Z^a)$, $f_{\text{EM}}^{(2)}(Z^a)$, each of which corresponds to an electromagnetic spinor in spacetime. We can then consider a homogeneity -2 function $f(Z^a)$, such that the combination

$$f_{-6}(Z^a) = \frac{f_{\text{EM}}^{(1)}(Z^a) f_{\text{EM}}^{(2)}(Z^a)}{f(Z^a)} \tag{5.83}$$

has homogeneity -6, where subscripts denote spins. The quantity $f_{-6}(Z^a)$ will correspond to a valid Weyl spinor in spacetime, due to the nature of the Penrose transform. Provided we are allowed to combine twistor functions in this way (a point we return to in the following), then our combination of functions in twistor space will necessarily correspond to some relationship between scalar, gauge, and gravity fields in spacetime. As argued in Refs. [138, 139], choosing particular functions in Eq. (5.83) leads precisely to the Weyl double copy of Eq. (5.49). To find the relevant functions, we may rely on a useful relationship between the singularities of a twistor function and the multiplicity of the principal spinors in the corresponding spacetime field. It turns out that, if a twistor function has only two distinct poles on the Riemann sphere X, then a pole of order m is associated with a $(2n - m + 1)$-fold degenerate principal spinor, in a field of spin n. The proof of this result is straightforward and relies on the Penrose transform [148] (see Ref. [139] for a recent review). Given that the Weyl double copy applies to type D solutions, this means that we wish to consider twistor functions of the general form

$$f_n(Z^a) = \left[Q_{ab} Z^a Z^b \right]^{-(n+1)}, \qquad (5.84)$$

where Q_{ab} is a constant dual twistor. The denominator contains a quadratic form in Z^a, which will have a pair of distinct poles in general, each of order $(n + 1)$. Thus, the corresponding spacetime fields will have a pair of distinct principal spinors, with multiplicities 1 and 2 for the electromagnetic and gravity cases, respectively. This is precisely what one expects for a type D solution that is obtained from a non-mixed double copy. To carry out the Penrose transform, we use the parametrisation of Eq. (5.79), noting that

$$d\pi_{E'} = (0, d\xi), \quad \lambda_{E'} d\lambda^{E'} = \epsilon^{A'B'} d\pi_{B'} \pi_{A'} = d\xi. \qquad (5.85)$$

Upon applying the incidence relation, the function of Eq. (5.84) will take the form

$$f_n(\xi, x) = \frac{N(x)}{[(\xi - \xi_1)(\xi - \xi_2)]^{n+1}}, \qquad (5.86)$$

for some function $N(x)$, where the parameters $\xi_{1,2}$ denote the positions of the singularities of $f(Z^a)$, in the ξ variable. The scalar case of Eq. (5.76) will now be

$$\phi = \frac{1}{2\pi i} \oint_\Gamma d\xi \frac{N(x)}{(\xi - \xi_1)(\xi - \xi_2)}, \qquad (5.87)$$

where we have carried out the contour integral by enclosing the pole at $\xi = \xi_1$ (n.b. we would get the same answer upon choosing $\xi = \xi_2$).

Now let us examine the spin-1 case:

$$\phi_{A'B'} = \frac{N^2(x)}{2\pi i} \oint_\Gamma d\xi \frac{(1,\xi)_{A'}(1,\xi)_{B'}}{(\xi - \xi_1)^2 (\xi - \xi_2)^2}. \tag{5.88}$$

We can again evaluate the integral by enclosing the (double) pole at $\xi = \xi_1$, upon recalling that the residue associated with an nth-order pole of a complex function $f(z)$ at $z = c$ is given by

$$\text{res}(f, c) = \frac{1}{(n-1)!} \lim_{z \to c} \frac{d^{n-1}}{dz^{n-1}}[(z-c)^n f(z)]. \tag{5.89}$$

Carrying out the integrals in Eq. (5.88) separately for each component, one ultimately finds (see e.g. Ref. [139] for more details)

$$\phi_{A'B'} = -\frac{2N^2(x)}{(\xi_1 - \xi_2)^3} \alpha_{(A'} \beta_{B')}. \tag{5.90}$$

Similarly, for the spin-2 case, one finds [139]

$$\phi_{A'B'C'D'} = \frac{6N^3(x)}{(\xi_1 - \xi_2)^5} \alpha_{(A'} \beta_{B'} \alpha_{C'} \beta_{D')}. \tag{5.91}$$

Converting notation to match Eq. (5.49) by setting

$$\phi_{A'B'C'D'}(x) = \Psi_{A'B'C'D'}(x), \quad \phi = \bar{S}(x),$$

we find that the fields of Eqs. (5.87), (5.90), and (5.91) indeed satisfy the Weyl double copy, where one may absorb an overall numerical factor into the scalar field.[10] To fully derive the Weyl double copy of Ref. [124], however, we must show that all possible vacuum type D solutions in gravity emerge from the above construction. That this is so follows from known results in the twistor literature [154], stating that all such solutions are indeed captured by the set of all possible constants Q_{ab} in Eq. (5.84).

A concrete example of the earlier general discussion is provided by the Schwarzschild solution, whose single copy has been discussed

---

[10] Recall that the Weyl double copy is itself only defined up to an overall numerical factor.

using the Kerr–Schild approach in the preceding chapter. The relevant quadratic form in that case is [155]

$$Q_{ab} = \begin{pmatrix} 0 & 0 & 0 & -1 \\ 0 & 0 & 1 & 0 \\ 0 & 1 & 0 & 0 \\ -1 & 0 & 0 & 0 \end{pmatrix}, \tag{5.92}$$

leading to the scalar field

$$\phi = \frac{i}{r\sqrt{2}} \propto \frac{1}{r} \tag{5.93}$$

as required, and the principal spinors

$$\alpha_{A'} = \left(1, \frac{x - iy}{z + r}\right), \quad \beta_{A'} = \left(1, \frac{x - iy}{z - r}\right). \tag{5.94}$$

Note that these do not immediately equal the principal spinors given for the Schwarzschild solution (obtainable from Kerr as $a \to 0$) in Eq. (5.52). However, all that is required is that the results match up to a simultaneous $\mathrm{SL}(2, \mathbb{C})$ transformation of both spinors.

The twistor double copy provides a derivation of the Weyl double copy whilst also providing a way of visualising properties of the latter geometrically. In particular, we see that the principal spinors of a spacetime field at a given point are associated with poles on the corresponding Riemann sphere X in twistor space. Due to the nature of the quadratic form of Eq. (5.84) needed to derive the type D Weyl double copy, these same poles are present in the twistor functions associated with the scalar, gauge, and gravity fields. This in turn gives a meaning to the "inverse zeroth copy" in Figure 1.2: given the twistor function associated with a particular scalar field, we can use its poles to dictate the principal spinors associated with the corresponding gauge and gravity solutions. This is in contrast to the Kerr–Schild double copy: if we have a scalar field in that approach, it is not at all clear what we have to dress it with to get a gauge or gravity solution. The twistor double copy also provides useful insight into why the Weyl double copy is local in position space, when one expects (following the BCJ double copy for scattering amplitudes) that it should be local in momentum space. We return to this point in the following.

In Ref. [139], the twistor double copy was also used to generate examples of Weyl double copies for gravity solutions whose Petrov type is other than D or N (n.b. the type N case can be viewed as a special case of type D). Given that the pattern of principal spinors corresponds to the locations of poles of twistor space functions, one can simply combine twistor functions with the desired properties to achieve a given Petrov type. The examples in Ref. [139] used, as building blocks, a set of twistor functions known as *elementary states* that first arose in the twistor literature as alternatives to plane waves for incoming and outgoing particles. These consist of simple ratios of rational factors e.g.

$$f(Z^a) = \frac{(C_c Z^c)^\gamma (D_d Z^d)^\delta}{(A_a Z^a)^\alpha (B_b Z^b)^\beta}, \tag{5.95}$$

where A_a, B_b, etc. are constant dual twistors. A novel physical interpretation of the corresponding fields in gauge theory and gravity has been provided by Refs. [156–160]. They correspond to field configurations in which the field lines are knotted together to create a nontrivial topology, where the linking is directly related to the various parameters α, β, etc. appearing in Eq. (5.95). The Weyl double-copy properties of such solutions were already addressed in Ref. [134], and the twistor approach described here extends this further. Note, however, that there are subtleties in how to interpret the Weyl double copy for non-type-D solutions. Unlike type D solutions, it is not true that solutions of general Petrov type automatically linearise the Einstein equations. Thus, the double copy one obtains can only be interpreted in the linearised theory and is no longer an exact statement. Furthermore, if more than one pole is present in twistor space, it is no longer true that the principal spinors of the gravity field are simply related to those of the electromagnetic field, and it can be more convenient to think about replacing Eq. (5.47) and (5.49) with a sum of terms, each of which has a simple double-copy interpretation. What the twistor methods thus confirm is that exact position-space double copies are very restricted in nature.

In the previous chapter, we saw that the Kerr–Schild double copy appears to extend to arbitrary conformally flat metrics. The twistor approach can explain this fact: as we saw earlier, the inner product for (dual) twistors is conformally invariant. Hence, the twistor space double copy for a given set of solutions should apply to all

gravity solutions in spacetime that are related by conformal transformations of the underlying Minkowski spacetime.[11] It seems, then,
that the twistor double copy is a useful method for elucidating the
scope of exact position-space double copies and clarifying their origin
and conceptual structure. However, up to now we have completely
ignored the issue of whether the product of twistor "functions"
appearing in Eq. (5.83) is well defined at all. Let us now address
this.

5.5.5   *How well defined is the twistor double copy?*

As is clear in Eq. (5.83), the procedure required in twistor space to
obtain the Weyl double copy in position space involves multiplying
together twistor functions. However, this raises a potentially serious
problem, given that each quantity $f(Z^a)$ is not a function but a
representative of a cohomology class, as discussed earlier. In more
pedestrian terms, we are supposed to be able to redefine each twistor
function according to the equivalence transformations of Eq. (5.80),
and yet doing so *before* forming the product of Eq. (5.83) will not
lead to the same gravity solution in general. The problem can be
traced to the fact that the combination in Eq. (5.83) is non-linear,
and indeed this non-linearity cannot be avoided given that the Weyl
double copy in position space is itself non-linear. How, then, are we
to reconcile the freedom of Eq. (5.80) with the twistor double copy
and thus provide a rigorous interpretation of the latter?

By far the simplest way of resolving this issue is to assume that the
twistor double copy of Eq. (5.83) indeed holds but that it demands
particular representatives of each cohomology class appearing on the
right-hand side. Indeed, this is analogous to how the BCJ double
copy for scattering amplitudes is only manifest in certain generalised
gauges, which require a particular choice of gauge-fixing and/or a
field redefinition (see Chapter 3). It may then be that a more general twistor double copy can be written down for general cohomology
representatives, such that this reduces to Eq. (5.83) in certain circumstances. This is all very well, but our insight is useless unless

---

[11]Strictly speaking, we must consider only those twistor functions that do not
involve the so-called *infinity twistors* that break conformal invariance [148].

we can find a particular method for systematically identifying the "special" twistor-space representatives of those gauge and gravity solutions that we wish to relate by the double copy.

The first paper to address this was Ref. [161], which examined gauge and gravity solutions involving radiation that propagates out to infinity. The authors used established twistor methods [162] to show that conditions at infinity could be imposed in order to pick out certain representatives of scalar, gauge, and gravity solutions, leading to a well-defined double copy. A different approach was used in Ref. [140], which used a known formulation of twistor theory in Euclidean rather than Minkowski space [163] – and using the mathematical language of *differential forms* – to show that there is a certain "minimal" choice of twistor representative for fields entering the double copy, such that a simple product in twistor space is indeed obtained. Unfortunately, however, it is not at all obvious how the approaches described in Refs. [140, 161] are related to each other nor to the original twistor double copy of Refs. [138, 139]!

A much firmer motivation for the twistor double copy was provided by Ref. [164]. It is known that classical solutions can be obtained from a certain inverse Fourier transform of scattering amplitudes in momentum space[12]: an example procedure for doing so is that of Ref. [165], which we revisit in Chapter 7. Reference [166] cunningly split this inverse Fourier transform into two stages, where the first turns a momentum-space amplitude into a "function" in twistor space. The second stage then turns out to be precisely the Penrose transform from twistor to position space. Reference [164] took the known forms of scattering amplitudes for a family of classical solutions (including the Schwarzschild and Kerr black holes) and explicitly demonstrated that the BCJ double copy for scattering amplitudes, the twistor double copy, and the Weyl double copy in position space amount to the same thing, albeit related by a precise set of integral transforms. The transform from momentum to twistor space picks out a special cohomology representative associated with each solution, and this turns out to be precisely the representative

---

[12]The relation between amplitudes and classical solutions requires analytic continuation of spacetime coordinates away from Minkowski space to a spacetime in $(2, 2)$ signature. References [143, 144] spell this out in detail.

expressed by Eq. (5.84)! This immediately puts both the twistor and Weyl double copies on a much firmer footing, and Ref. [164] confirmed that exact position-space double copies are highly unusual but rigorous where they apply.

A puzzle regarding both the Kerr–Schild and Weyl double copies is why they are local in position space, whereas the BCJ double copy for scattering amplitudes is local in momentum space. As shown in Ref. [164], it is only for very special solutions (such as those sourced by certain scattering amplitudes) that one obtains a local product in twistor space. This then translates into a local statement in position space, given that the Weyl double copy is a statement about how the principal spinors of a gravity solution are obtained from those of gauge theory solutions. The principal spinors of a field are dependent on spacetime position, and thus the Weyl double copy must act point-by-point in spacetime. However, there is also a non-locality associated with the Weyl double copy: it is a statement about *directions*, which are associated with extended objects (lines) in Minkowski spacetime. Thus, there is ultimately no conflict between the locality of the Weyl double copy in spacetime and the non-local nature of the map from twistor space to position space.

5.6 Equivalence of the Kerr–Schild and Weyl Double Copies

Having discussed the Weyl double copy at length, let us now see how it can be related to the Kerr–Schild double copy of the previous chapter. Earlier, we saw that spinor fields can be classified in terms of their principal spinors. If we return to standard (real) Minkowski space from the complexified version considered earlier, then we can associate a null vector with each principal spinor as follows:

$$k^\mu(x) = \alpha_{A'}(x)\bar{\alpha}_A(x)\sigma^\mu_{AA'},   \qquad (5.96)$$

where the bar denotes complex conjugation. This is nothing more than the statement that the spinorial translation of a null vector factorises into the outer product of a spinor and its complex conjugate, and that we may convert to the tensor language by contracting with the appropriate Infeld–van der Waerden symbol. Given that the spinors depend on spacetime position in general, so will the null

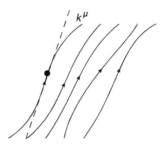

Fig. 5.3. A principal spinor defines a null direction at each point in spacetime that is tangent to a *null congruence*, namely a set of field lines filling all space, all of whose tangent vectors are null.

vector k^μ. It thus defines a *null congruence* in spacetime, namely a set of spacetime curves, all of whose tangent vectors are null. This is depicted in Figure 5.3, and our twistor derivation of the Weyl double copy in fact allows us to say more than this. A result known as the *Kerr theorem* [148] states that the vanishing of a function $g(Z^a)$ in twistor space is associated with a *null, shear-free, geodesic congruence* in spacetime.[13] In simple terms and for our purposes: the Kerr theorem directly implies that the vector field k^μ satisfies the Kerr–Schild conditions of Eqs. (4.5) and (4.7)! We thus have that the types of solutions generated by the Weyl and Kerr–Schild double copies are the same. For type D solutions, there are two distinct principal spinors and thus a pair of null geodesic (shear-free) congruences.

An explicit example of this comparison is provided by the Schwarzschild solution, whose principal spinors have been given in Eq. (5.94). We can thus form two Kerr–Schild vectors according to Eq. (5.96), and these turn out to be [139]

$$k_\mu^{(1,2)} \propto (1, \pm 1, 0, 0) \qquad (5.97)$$

in spherical polar coordinates (t, r, θ, ϕ). One of these is the Kerr–Schild vector that we used in Chapter 4, and the other is an alternative choice that we could have used instead, in that it is equivalent to the standard choice by a gauge or coordinate transformation in gauge theory/gravity.

---

[13]The term "shear-free" here describes a lack of distortion of small shapes which are perpendicular to the vector fields, as they are transported along the field lines.

In the Weyl double copy, one introduces separate scalar fields $S(x)$ and $\bar{S}(x)$ for the (anti-)self-dual part of the field strength spinor. For real solutions, these are related by complex conjugation so that the overall zeroth copy of a given solution is

$$\phi(x) = S(x) + \bar{S}(x). \tag{5.98}$$

As a non-trivial example, we may take the Kerr scalar field of Eq. (5.55), which yields

$$S(x) + \bar{S}(x) = -\frac{Q^2}{6M}\left[\frac{1}{r - iaX} + \frac{1}{r + iaX}\right] = \frac{Q^2}{3M}\frac{r}{r^2 + a^2 X^2}.$$

Substituting

$$X = \cos\theta = \frac{z}{r},$$

one reproduces (up to an allowable constant factor) the scalar function in the Kerr–Schild approach of Eq. (4.26).

5.7   Sources in the Weyl Double Copy

So far, our discussion of the Weyl double copy has been restricted to vacuum solutions of the equations of motion. References [167, 168] have shown how the Weyl double copy may be generalised to include the effects of non-trivial source terms for scalar, gauge, or gravitational fields. The basic idea is to replace Eq. (5.47) with the following:

$$\Psi_{ABCD} = \sum_{n=1}^{m}\frac{1}{S_{(n)}}\Phi_{(AB}^{(n)}\Phi_{CD)}^{(n)}, \tag{5.99}$$

and likewise for the dual Weyl spinor. We now have a tower of electromagnetic spinors $\Phi_{AB}^{(n)}$ and scalar fields $S^{(n)}(x)$, where the $n = 1$ term is taken to correspond to the conventional (vacuum) double copy. In the remaining terms, each electromagnetic (dual) spinor corresponds to a tensorial field strength $F_{\mu\nu}^{(n)}$ satisfying a non-vacuum

Maxwell equation:

$$\partial^\nu F^{(n)}_{\mu\nu} = J^{(n)}_\mu, \tag{5.100}$$

where $J^{(n)}_\mu$ is an appropriate source current. Likewise, each scalar field satisfies a non-vacuum Klein–Gordon equation:

$$\partial^2 S^{(n)} = \rho^{(n)}_S, \tag{5.101}$$

where $\rho^{(n)}_S$ is a charge density. Equation (5.99) thus corresponds to splitting a given gravity solution into m distinct pieces, each of which has a well-defined single and zeroth copy. One then needs a prescription for how to choose each distinct piece i.e. how to identify each individual single and zeroth copy $\Phi^{(n)}_{AB}$ and $S^{(n)}$. As argued in Refs. [167, 168], this can be done by requiring that the sources for $\Phi^{(n)}_{AB}$ and $S^{(n)}$ match up in some sensible physical way with their gravitational counterpart (i.e. a non-zero contribution to the Ricci tensor). Many examples are provided by the authors, and we mention a couple here to illustrate the idea. First, one may consider a toy black hole solution with line element

$$ds^2 = -a(r)dt^2 + \frac{1}{a(r)}dr^2 + r^2 d\Omega^2, \tag{5.102}$$

where t and r are time and radial coordinates and $d\Omega$ the conventional unit of solid angle. For the function $a(r)$, we may assume the general form

$$a(r) = 1 + \sum_{n=1}^{m} a_n r^{-n}, \tag{5.103}$$

and the Weyl spinor for this solution turns out to be

$$\Psi_{ABCD} = \sum_{n=1}^{m} \frac{(n+1)(n+2)}{2} \frac{a_n}{r^{n+2}} o_{(A}o_B \iota_C \iota_{D)}, \tag{5.104}$$

where

$$o_A = \frac{1}{\sqrt{2}}(1,1), \quad \iota_A = \frac{1}{\sqrt{2}}(1,-1). \tag{5.105}$$

We may then identify the individual scalar fields and electromagnetic spinors for each term in Eq. (5.104) as follows:

$$S^{(n)} \propto \frac{1}{r^n}, \quad \Phi^{(n)}_{AB} \propto \frac{1}{r^{n+1}} o_{(A} \iota_{B)}. \tag{5.106}$$

Plugging all solutions into their respective field equations, one finds that the scalar charge density $\rho^{(n)}_S$, electromagnetic charge density $\rho^{(n)}_e$, and gravitational energy density $\rho^{(n)}_{\text{grav.}}$ satisfy

$$\rho^{(n)}_S \propto \rho^{(n)}_e \propto \rho^{(n)}_{\text{grav.}} \propto \frac{1}{r^{n+2}}, \tag{5.107}$$

thus validating the "physical" identification of each term in the generalised Weyl double copy.

Another application of this formalism – spelled out in detail in Ref. [168] – is in describing solutions of gravity coupled to electromagnetism (*Einstein–Maxwell theory*) as special cases of Eq. (5.99), with $m = 2$. That is, one may write

$$\Psi_{ABCD} = \frac{1}{S^{(1)}} \Phi^{(1)}_{(AB} \Phi^{(1)}_{CD)} + \frac{1}{S^{(2)}} \Phi^{(2)}_{(AB} \Phi^{(2)}_{CD)}, \tag{5.108}$$

where the corresponding gauge and scalar fields $A^{(1)}_\mu$ and $S^{(1)}$ satisfy a vacuum equation, but where $A^{(2)}_\mu$ and $S^{(2)}$ require source currents/charges that match up in a physically sensible way with the corresponding gravity result. A test case is the canonical *Kerr–Newman black hole*, which is an electromagnetically charged analogue of the Kerr black hole considered here previously. Its Weyl spinor may be written as

$$\Psi_{ABCD} = 6 \left(-\frac{M}{(r + ia\cos\theta)^3} + \frac{Q^2}{(r + ia\cos\theta)^3(r - ia\cos\theta)} \right)$$
$$\times o_{(A} o_B \iota_C \iota_{D)}, \tag{5.109}$$

where the first and second terms match up with the corresponding terms in Eq. (5.108), upon choosing

$$S^{(1)} \propto \frac{1}{r + ia\cos\theta}, \quad \Phi^{(1)}_{AB} \propto \frac{1}{(r + ia\cos\theta)^2} o_{(A} \iota_{B)} \tag{5.110}$$

and

$$S^{(2)} \propto \frac{1}{(r + ia\cos\theta)(r - ia\cos\theta)},$$

$$\Phi_{AB}^{(2)} \propto \frac{1}{(r + ia\cos\theta)^2(r - ia\cos\theta)} o_{(A} \iota_{B)}. \qquad (5.111)$$

The fields in Eq. (5.111) are such that the scalar charge, electromagnetic charge, and gravitational energy densities satisfy

$$\rho_S^{(2)} \propto \rho_e^{(2)} \propto \rho_{\text{grav.}}^{(2)} \propto \frac{(r^2 + a^2) + a^2\sin^2\theta}{\rho^6}. \qquad (5.112)$$

There is also a current in the gauge theory (due to the rotation of the black hole) that matches up with a gravitational counterpart.

Given that the fields considered in this section are no longer vacuum solutions, there is a non-zero Ricci tensor in the gravity theory, and thus the *Ricci spinor* $\Phi_{ABC'D'}$ appearing in Eq. (5.40) will be non-zero, in addition to the Weyl spinor Ψ_{ABCD}. Reference [168] argues that the Ricci spinor may also be written in a double-copy form e.g. as

$$\Phi_{ABC'D'} = \frac{1}{S^{(2)}} \Phi_{AB}^{(2)} \Phi_{C'D'}^{(2)} \qquad (5.113)$$

in the earlier notation, for those solutions whose Riemann tensor vanishes at infinity. Other (but related) formulae have been proposed in the context of $\mathcal{N} = 0$ supergravity [169], using the twistor methods described earlier.

5.8 The Convolutional Double Copy

To close this chapter, we examine another widely studied double-copy procedure for classical solutions that is complementary to the Kerr–Schild and Weyl double copies considered thus far and works for a broader class of spacetimes in principle. It takes as its starting point that the BCJ double copy for scattering amplitudes involves products in momentum space. Given that position and momentum space are related by (inverse) Fourier transformation, one expects that products in momentum space should become *convolutions* in position space, where the convolution of two functions $f(x)$ and $g(x)$ of 4-vector position is defined by

$$[f \star g](x) = \int d^4 y f(y) g(x - y). \tag{5.114}$$

We should therefore be able to write double-copy formulae for arbitrary gravity solutions in terms of convolutions of gauge and scalar fields, and Ref. [170] presents just such an approach, which we refer to as the *convolutional double copy*. It was used in Refs. [171–178] to construct a catalogue of double copies in various exotic generalisations of gauge theory and gravity. But it has also become increasingly important in addressing key conceptual questions regarding the double copy.

As we saw in Chapter 3, the double copy for amplitudes is only manifest if one chooses a particular (generalised) gauge on both the gauge and gravity sides of the correspondence. This property is shared by the Kerr–Schild double copy: not only does this only work for very particular solutions, but it also requires the choice of very special (Kerr–Schild) coordinates and thus a particular choice of "gauge" in gravity. The Weyl double copy is more gauge-independent but is after all restricted to solutions of linearised gravity and gauge theory. It becomes exact only for those special solutions that happen to linearise the field equations, such as the vacuum Petrov type D solutions we studied earlier. However, the convolutional approach is much more all-encompassing, in that it can be chosen to relate fields in *arbitrary* gauges. To see how this works, let us recall from Chapter 3 that in general gauges, a non-abelian gauge field will have an unphysical longitudinal polarisation state, in addition to the two transverse polarisation states expected for a massless particle. The unphysical state will contribute to scattering amplitudes beyond tree level, and one may introduce additional fields called *ghosts*, whose job is to explicitly subtract the unwanted degrees of freedom and leave a gauge-invariant result. We can then write a complete set of equations of motion for Yang–Mills theory in an arbitrary gauge, involving the ghost fields in addition to the gluon and any matter fields that may be present.

Similar to the gauge theory case, arbitrary "gauges" (choices of coordinates) in gravity will produce a graviton field that is not transverse or traceless and thus carries additional unphysical degrees of freedom beyond the two transverse polarisation states expected for a physical graviton. One may again introduce ghost fields, and there are more of them given that there are more degrees of freedom that need subtracting. However, a completely systematic procedure exists for determining the ghost equations of motion for a given gauge choice [179–184] that generalises previously known results in gauge theory. In the latter, one may show that there is a so-called *BRST symmetry* [185–188] obeyed by the equations defining the theory. It can be described by the action of an abstract operator Q that acts on the various fields to transform them into other fields as follows:

$$Q\mathbf{A}_\mu = \partial_\mu \mathbf{c}, \quad Q\mathbf{c} = 0, \quad Q\bar{\mathbf{c}} = G[\mathbf{A}_\mu], \qquad (5.115)$$

where $\mathbf{A}_\mu(x)$ is the gauge field, $\mathbf{c}(x)$ the ghost field, and $\bar{\mathbf{c}}(x)$ the anti-ghost field (i.e. the antimatter counterpart of $\mathbf{c}(x)$). Furthermore, $G[\mathbf{A}_\mu]$ is the condition one uses to fix a particular gauge e.g.

$$G[\mathbf{A}_\mu] = \partial \cdot \mathbf{A} \qquad (5.116)$$

would correspond to the gauge-fixing condition $\partial_\mu A^{\mu a} = 0$ for each component $A^{\mu a}$ of the gauge field, which is commonly known as the *Lorenz gauge*. The requirement that the equations of motion for all fields should be gauge-invariant then amounts to the statement that their form should be unchanged under action of the BRST operator Q.

A similar story can be told in gravity, where we focus on the double copy of pure Yang–Mills theory for completeness. As we saw in Chapter 3, the double copy of YM theory is not pure general relativity but includes two additional fields, known as the axion and dilaton. Nevertheless, one may introduce a BRST-like operator Q which acts on all fields. For example, its action on the graviton $h_{\mu\nu}$ and dilaton field φ can be written as

$$Qh_{\mu\nu} = \partial_\mu c_\nu + \partial_\nu c_\mu, \quad Q\varphi = 0, \qquad (5.117)$$

where $c_\mu(x)$ is an appropriate ghost field, which itself satisfies

$$Qc_\mu = 0, \quad Q\bar{c}_\mu = G_\mu[h_{\mu\nu}, \varphi] \qquad (5.118)$$

together with its antighost. Here, G_μ is the gravitational gauge-fixing condition, for which a common example is the *de Donder gauge*

$$G_\mu[h_{\mu\nu}, \varphi] = \partial^\nu \bar{h}_{\mu\nu}, \quad \bar{h}_{\mu\nu} = h_{\mu\nu} - \frac{h}{2}\eta_{\mu\nu}, \quad h \equiv h^\alpha_\alpha. \quad (5.119)$$

References [189, 190] introduced a systematic procedure for combining the convolutional double copy with the BRST formalism for gauge invariance. First, one writes a general dictionary expressing fields in a gravity theory in terms of convolutions of fields in two single-copy gauge theories, including ghosts on both sides. As an example, for a gravity solution with non-zero graviton and dilaton, one may write the general ansatz [191]

$$h_{\mu\nu} = 2\mathbf{A}_\mu \circ \mathbf{A}_\nu + \eta_{\mu\nu}\left(a_1\mathbf{A}^\rho \circ \mathbf{A}_\rho + a_2\mathbf{c} \circ \bar{\mathbf{c}}\right),$$

$$\varphi = a_3\mathbf{A}^\rho \circ \mathbf{A}_\rho + a_4\mathbf{c} \circ \bar{\mathbf{c}}. \quad (5.120)$$

Following e.g. Ref. [191], we have introduced the *circle product*

$$\mathbf{A}_\mu \circ \mathbf{A}_\nu = A^a_\mu \star \Lambda^{aa'} \star A^{a'}_\nu, \quad \mathbf{c}_\mu \circ \bar{\mathbf{c}}_\nu = c^a_\mu \star \Lambda^{aa'} \star \bar{\mathbf{c}}^{a'}_\nu, \quad (5.121)$$

where $\{a_i, \xi\}$ are constants and $\Lambda^{aa'}$ is a so-called *spectator field*, whose job is to contract the colour indices of the gauge fields and ghosts, leaving a result that is free of colour indices. The first equation of Eq. (5.121) shows us that the spectator field acts like the *inverse* of the biadjoint field in the Kerr–Schild and Weyl double copies, in that it must also remove that part of the gauge field that is counted twice in forming the product of gauge fields.

Equation (5.120) constitutes a general dictionary between gauge and gravity fields, where we have yet to fix a gauge on each side of the double copy. We can achieve the latter by applying the BRST operator Q to Eq. (5.120) and assuming that this is the *same* BRST operator that acts in the gauge theory. By equating the result to the general form of the BRST transformations in gravity, we get consistency relations that can be used to fix the arbitrary parameters in the dictionary. For example, from Eq. (5.120), one finds [191]

$$Qh_{\mu\nu} = 4\partial_{(\mu}\mathbf{c} \circ \mathbf{A}_{\nu)} + \eta_{\mu\nu}\left(2a_1 - a_2\right)\partial^\rho \mathbf{c} \circ \mathbf{A}_\rho, \quad (5.122)$$

where the brackets in the first term denote symmetrisation over indices as usual. Comparing with Eq. (5.117) yields the constraint

$$a_2 = 2a_1, \qquad (5.123)$$

as well as the identifications[14]

$$c_\mu = 2\mathbf{c} \circ \mathbf{A}_\nu, \quad \bar{c}_\mu = 2\bar{\mathbf{c}} \circ \mathbf{A}_\nu. \qquad (5.124)$$

Going further, one finds

$$Q\bar{c}_\mu = 2\left[\partial^\rho \mathbf{A}_\rho \circ \mathbf{A}_\mu + 2\partial_\mu(\mathbf{c} \circ \bar{\mathbf{c}})\right]$$

$$= \left[\partial^\rho \bar{h}_{\rho\mu} + \frac{(1 + a_1)}{a_3}\partial_\mu \varphi\right],$$

where we have used Eqs. (5.120) and (5.123). We may now compare this with Eq. (5.118), by implementing the particular gauge condition of Eq. (5.119). In so doing, we fix

$$a_1 = -1 \qquad (5.125)$$

and thus a specific dictionary between gauge/gravity fields in our chosen gauges. Such analyses can be extended to a wide variety of theories and gauges, where any remaining parameters can be associated with residual gauge transformations compatible with a given gauge-fixing condition. Thus, the convolutional double copy gets closer than any other approach in being able to precisely relate the gauge transformations of non-abelian gauge theory with the coordinate transformations (and field redefinitions) of gravity, thereby showing that the double copy is a property of the "complete" theories in some sense. Once a dictionary has been fixed for particular gauge choices, one may examine specific solutions, and even look for new gravity solutions by combining gauge theory results: see e.g. Ref. [192] for an example. Traditionally, the convolutional double copy has been restricted to linear order in perturbation theory

---

[14]In deriving Eq. (5.124), it is useful to note the property $\partial_\mu[f(x) \circ g(x)] = [\partial_\mu f(x)] \circ g(x)$, which is inherited from a similar property of the convolution operation.

only, which is not a major restriction in that its competing classical double copies are themselves only exact for solutions that linearise the field equations. Progress towards extending the convolution approach to higher orders in perturbation theory has been made in Ref. [193].

Chapter 6

Towards a Non-Perturbative
Double Copy

In the previous two chapters, we have examined various double-copy procedures for classical solutions, which indeed demonstrate that the scheme of Figure 1.2 is a lot more general than being merely restricted to scattering amplitudes in each theory. All of the formalisms we considered for classical solutions were ultimately restricted to the linearised equations of each field theory, due to our wishing to relate *exact* solutions. It is already known that one can describe more general solutions if one relaxes this condition, where the price one pays is having to work order-by-order in perturbation theory for classical solutions, as one does for (quantum) scattering amplitudes. We describe such results in the following chapter, but it is first interesting to ask if there are any hints that the double copy may extend to properties that go *beyond* perturbation theory. Such non-perturbative insights are conceptually important in telling us quite how generally the double copy is meant to apply, but may also have more practical applications, in telling us how to find new gravity solutions.

In this chapter, we review a number of topics relating to non-perturbative aspects of the double copy, a subject which is generally less well-explored than perturbation theory. This is largely due to ignorance of how to proceed: in all other incarnations of the double copy, solutions of the linearised theory play a crucial role. In scattering amplitudes, for example, one must leave the denominators of individual cubic graphs untouched whilst replacing information in the numerator. In the Kerr–Schild, Weyl and convolutional double copies, one must leave untouched (or correct for) the presence of a scalar field that itself obeys a linear equation. When non-linear and/or non-perturbative situations are encountered, it is simply not known what the rules of the double copy might be, and this has led to a number of complementary studies in the research literature in recent years. Some of these are concerned with collecting useful theoretical data that might be used to eventually formulate a non-perturbative double copy, with a particular focus on solutions of biadjoint scalar field theory [194–196]. Other work has looked at how symmetries [37, 45, 135, 197, 198] or geometrical/topological properties [199–201] might match up. More recent work has argued that a fully non-perturbative realisation of the double copy may be given in special circumstances [202, 203]. Given how little is known, all approaches and insights are equally welcome!

6.1 Kinematic Algebras

In Section 3.1, we saw that the double copy for scattering amplitudes relied upon the intriguing property of *BCJ duality*: the kinematic numerators $\{n_i\}$ entering the amplitudes of Eqs. (3.23) and (3.34) obey similar relations to the colour factors $\{c_i\}$ in the gauge theory, namely the Jacobi identities that interrelate triplets of colour factors. The latter arise from the Lie algebra corresponding to the gauge symmetry of the theory, and this in turn leads us to believe that there must be a *kinematic algebra* that somehow mirrors the colour algebra. To date, a detailed understanding of what this kinematic algebra might be – and even whether it exists at all for general theories – has been missing. This is largely due to the fact that BCJ duality in an amplitude context is an intrinsically perturbative phenomenon that relies on first eliminating 4-gluon vertices in favour of cubic ones, as discussed in Section 3.3. This is possible order-by-order in perturbation theory, but does not give much clue as to whether there is some deeper underlying property, which in turn gives rise to the Jacobi relations between kinematic factors.

Remarkably, there are a few cases in which the kinematic algebra can be understood completely. The most well-known of these is the case of *self-dual Yang–Mills theory*, where one keeps only one polarisation state of the gauge field such that the field strength tensor is self-dual. One may do a similar thing in gravity, and the theory of a self-dual graviton is indeed the double copy of self-dual Yang–Mills theory. There is then a particular way of presenting these theories that makes the double-copy structure manifest [45]. First, one may define the coordinates

$$u = \frac{t - z}{\sqrt{2}}, \quad v = \frac{t + z}{\sqrt{2}}, \quad X = \frac{x + iy}{\sqrt{2}}, \quad Y = \frac{x - iy}{\sqrt{2}}, \qquad (6.1)$$

in terms of which the line element for Minkowski spacetime becomes

$$ds^2 = -2du\,dv + 2dX\,dY. \qquad (6.2)$$

One may then define the differential operator (in the (u, v, X, Y) coordinate system)

$$\hat{k}_\mu = -(0, \partial_Y, \partial_u, 0) \qquad (6.3)$$

and define a graviton field according to

$$h_{\mu\nu} = \hat{k}_\mu \hat{k}_\nu \phi, \qquad (6.4)$$

where ϕ is a scalar field. Substituting this into the Einstein equations, one finds that the field ϕ must satisfy

$$\partial^2 \phi - \frac{1}{2}(\hat{k}^\mu \hat{k}^\nu \phi)(\partial_\mu \partial_\nu \phi) = 0, \qquad (6.5)$$

which turns out to agree with the so-called *Plebanski equation* of self-dual gravity, first presented in Ref. [204]. We may make a similar ansatz for Yang–Mills theory, and the most general possibility is to define a gauge field

$$A_\mu^a = \hat{k}_\mu \Phi^a, \qquad (6.6)$$

where Φ^a is a scalar field with a colour index a. Substituting this ansatz into the Yang–Mills equations, one finds that the latter reduce to [89]

$$\partial^2 \Phi^a - \frac{1}{2} f^{abc}(\hat{k}^\mu \Phi^b)(\partial_\mu \Phi^c) = 0, \qquad (6.7)$$

which turns out to reproduce a known equation of motion for self-dual Yang–Mills theory [205]. We have here made certain choices for how to normalise the coupling constants, so as to match conventions with Ref. [203]. In the same notation, we may quote the field equation for biadjoint scalar theory, which is

$$\partial^2 \Phi^{aa'} - \frac{1}{2} f^{abc} \tilde{f}^{a'b'c'} \Phi^{bb'} \Phi^{cc'} = 0. \tag{6.8}$$

Looking at Eqs. (6.5), (6.7), and (6.8), we see that they are all at most quadratic in the relevant field, and this turns out to generate Feynman diagrams with at most cubic vertices. Thus, the above obstruction to understanding the non-perturbative nature of the kinematic algebra (i.e. that there are higher-than-cubic vertices in the theory) is removed. We can then work out what the kinematic algebra is. To do this, note that in moving from Eq. (6.5) to Eq. (6.7), a certain combination of derivative operators is replaced by a set of colour structure constants and that this gets repeated as we move to Eq. (6.8). It is then possible to identify the derivative operators themselves as being related to a set of "kinematic structure constants", associated with an infinite-dimensional Lie algebra. As it happens, this is easier to see in momentum space, so let us define the Fourier transform of a position-space field according to

$$\tilde{A}(p) = \int d^4x A(x) e^{ip \cdot x}, \quad A(x) = \int \frac{d^4p}{(2\pi)^4} \tilde{A}(p) e^{-ip \cdot x}. \tag{6.9}$$

We may then Fourier transform Eq. (6.7), whilst also writing the fields in the second term as inverse Fourier transforms. The result is

$$-p_1^2 \tilde{\Phi}^a(p_1) - \frac{1}{2} \sum_{p_2,p_3} F_{p_2p_3}{}^{p_1} f^{abc} \tilde{\Phi}^b(p_2) \tilde{\Phi}^c(p_3) = 0, \tag{6.10}$$

where we have defined the quantity

$$F_{p_2p_3}{}^{p_1} = \delta^{(4)}(p_1 - p_2 - p_3) X(p_2, p_3),$$
$$X(p_2, p_3) = -p_{2X}p_{3v} + p_{2v}p_{3X}, \tag{6.11}$$

as well as the momentum sum

$$\sum_p \equiv \int \frac{d^4p}{(2\pi)^4}. \tag{6.12}$$

Equation (6.10) has a highly suggestive form, involving a non-linear term that looks like the product of two different structure constants.

That the quantities $F_{p_2 p_3}{}^{p_1}$ are indeed associated with a Lie algebra can be seen as follows. Let us first define the generators

$$V_{p_i} = -(\hat{k}^\mu e^{ip_i \cdot x})\partial_\mu, \qquad (6.13)$$

whose interpretation we shall return to in the following. We can calculate the commutator of these generators by acting on an arbitrary test function $f(x)$, yielding

$$[V_{p_2}, V_{p_3}] = X(p_2, p_3)V_{p_2+p_3} = \sum_{p_1} F_{p_2 p_3}{}^{p_1} V_{p_1}, \qquad (6.14)$$

where we have recognised the action of the structure constant of Eq. (6.11). This tells us that the generators of Eq. (6.13) indeed form a Lie algebra[1] and that the quantity of Eq. (6.11) is the appropriate structure constant. This lends a particularly nice interpretation to the transition from the biadjoint scalar field equation of Eq. (6.8) to the self-dual Yang–Mills equation of (6.7): to obtain the latter from the former, one can simply replace one set of colour structure constants with the kinematic structure constants of Eq. (6.11). Furthermore, the fact that the non-linear term in the field equation involves an explicit product of structure constants means that whenever there is a Jacobi identity for one of the Lie algebras, the same will be true for the other. The self-dual sector is thus arguably the cleanest known case in which a gauge theory is seen to have a kinematic Lie algebra that mirrors the colour algebra. The double and zeroth copies can be seen as replacements of one kind of algebra with another: as is made clear in Ref. [45], the Plebanski equation of self-dual gravity can itself be shown to involve two copies of the kinematic structure constants.

Having now found the generators for the kinematic algebra of self-dual Yang–Mills theory, let us now interpret what they actually correspond to. To this end, let us introduce a function

$$V(x) = \int \frac{d^4 p}{(2\pi)^4} \tilde{V}(p) e^{ip \cdot x}, \qquad (6.15)$$

---

[1]There is also a Jacobi identity for the kinematic structure constants, as described in Ref. [45].

such that a general superposition of the generators in Eq. (6.13) takes the form

$$V^\mu(x)\partial_\mu, \quad V^\mu(x) = \hat{k}^\mu V. \qquad (6.16)$$

We may interpret $V^\mu(x)$ as a vector field in spacetime and then note that $V^\mu\partial_\mu$ at the point x generates an infinitesimal translation in the direction of V^μ. Thus, Eq. (6.16) considered over all spacetime generates a simultaneous translation along all the field lines of the vector field $V^\mu(x)$. This is called a *diffeomorphism* and is restricted further by the special form of $\hat{k}^\mu$. From its definition in Eq. (6.3), one may show that

$$\partial_\mu \hat{k}^\mu = 0, \qquad (6.17)$$

and a standard result in differential geometry (see e.g. Ref. [206]) states that a diffeomorphism is volume-preserving provided its associated vector field satisfies

$$\partial_\mu V^\mu = 0. \qquad (6.18)$$

That is, if we take a small shape and watch how it becomes deformed upon transporting it along the field lines using the diffeomorphism, the volume of this shape will not change. Our vector field of Eq. (6.16) indeed satisfies this criterion by virtue of Eq. (6.17). More than this, Eq. (6.3) implies that

$$V^\mu\partial_\mu = V^u\partial_u + V^Y\partial_Y, \qquad (6.19)$$

i.e. that no other derivatives appear on the right-hand side. The diffeomorphisms generated by V^μ thus take place only in the infinite family of planes associated with fixed values of v and X. This is shown in Figure 6.1, and we can clearly consider the diffeomorphisms as acting in each plane independently. The above remarks then imply that the diffeomorphisms will be *area-preserving* in each plane, and thus we conclude that the kinematic Lie algebra of self-dual Yang–Mills theory is one of area-preserving diffeomorphisms, as first noted in Ref. [45]. As argued in Refs. [89, 203], any vector operator satisfying Eq. (6.17) and $\hat{k}^2 = 0$, when substituted in the ansätze of Eqs. (6.4) and (6.6), will yield the self-dual field equations of (6.5) and (6.7). One is thus free to rotate the coordinate system,

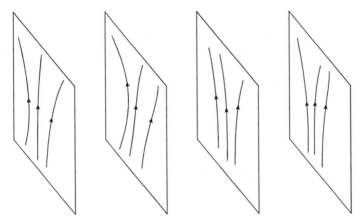

Fig. 6.1. The kinematic algebra of self-dual Yang–Mills theory consists of diffeomorphisms (simultaneous translations) along field lines contained in the infinite family of planes at fixed u and X values. These diffeomorphisms are area-preserving, and one may choose any set of α-planes by rotating the coordinates.

and to consider the kinematic algebra as generating area-preserving diffeomorphisms in any family of α-planes, to use the terminology of the previous chapter.

There are other situations where a similar kinematic algebra appears. For example, Ref. [207] examined generalisations of self-dual Yang–Mills and related theories, showing that other field equations can be written in terms of kinematic structure constants, where the area-preserving property is relaxed. Reference [202] looked at a novel family of theories in two spacetime dimensions, whose kinematic algebras also consist of area-preserving diffeomorphisms. A known result [208] states that the Lie group SU(N), for $N \to \infty$, becomes equivalent to the group of diffeomorphisms of a torus. Reference [202] used this to argue that certain 2d analogues of SU(N)×SU(N) biadjoint scalar theory, SU(N) gauge theory, and self-dual gravity are completely equivalent as $N \to \infty$, thus providing some sort of motivation for the replacement of colour by kinematic algebras. More generally, Ref. [209] examined a three-dimensional variant of Yang–Mills theory known as *Chern–Simons theory* and showed that the kinematic algebra in that case could be fully understood as a volume-preserving diffeomorphism algebra.

Above, we have seen that at the level of equations of motion, the double copy operates by replacing the structure constants of one

Lie algebra with those of another. We may also ask how it operates for the fields themselves i.e. for given solutions of the equations of motion. This has been clarified in Ref. [203], as follows. First, let us recall that the full biadjoint scalar field consists of contracting the component fields $\Phi^{aa'}$ with generators $\mathbf{T}^a$ and $\tilde{\mathbf{T}}^{a'}$ associated with each gauge group:

$$\Phi = \mathbf{T}^a \tilde{\mathbf{T}}^{a'} \Phi^{aa'}. \tag{6.20}$$

Likewise, we can think of the gauge field A^a_μ (at least at linearised level) as generating diffeomorphisms along its field lines, such that one may replace one of the colour generators appearing in Eq. (6.20) with the relevant diffeomorphism generator:

$$\mathbf{A}^\mu \partial_\mu = \mathbf{T}^a (\hat{k}^\mu \Phi^a) \partial_\mu. \tag{6.21}$$

Finally, we may think of the graviton field $h_{\mu\nu}$ as generating two sets of diffeomorphisms:

$$h^{\mu\nu} \partial_\mu \partial_\nu = (\hat{k}^\mu \hat{k}^\nu \phi) \partial_\mu \partial_\nu. \tag{6.22}$$

In summary, where the kinematic algebra is fully understood, one may double copy equations of motions by replacing structure constants and solutions by replacing generators. Various references have tried to generalise our understanding of kinematic algebras to other theories and/or special kinematic limits or to study higher-level mathematical structures for doing so [37, 46, 207, 210–226]. What some of these references make clear is that the kinematic algebra may not simply be a Lie algebra in general but some more complicated mathematical structure. For example, it may be the case that the Jacobi identity is only satisfied up to certain corrections which vanish on-shell and which can be formalised in terms of abstract brackets that generalise the simple commutator to be found in a Lie algebra. However, it is perhaps fair to say that at the time of writing, a detailed geometric understanding comparable to the above discussion remains elusive.

6.2 Exact Solutions of Biadjoint Scalar Field Theory

As well as looking for specific structures that may match up between different theories, one may also examine specific exact non-linear solutions and see if these can be matched up according to some double-copy procedure that goes beyond the double copies considered in previous chapters. A large number of exact solutions of Yang–Mills theory are known (see e.g. Refs. [227, 228] for reviews). However, it has not proved easy to match these up with gravity solutions. It may therefore be simpler to consider the left-hand part of Figure 1.2 and to try to match up exact non-linear solutions of biadjoint scalar field theory with those in pure Yang–Mills theory. Motivated by this, Refs. [194–196, 203] have started to accrue a catalogue of non-linear biadjoint scalar field solutions and also speculated how they might be related to gauge theory counterparts.

The biadjoint scalar field equation has been given in Eq. (3.36), and we may start by looking for field solutions with the simpler form

$$\Phi^{aa'}(x) = \chi^a(x)\xi^{a'}(x). \tag{6.23}$$

However, substituting this into Eq. (3.36) shows that the non-linear term vanishes so that one can only obtain non-linear solutions if the two different types of colour charge are inextricably linked. This is itself an interesting conclusion, and the next simplest solutions to look for are those which are both static and spherically symmetric, for which Eq. (3.36) reduces to

$$\nabla^2 \Phi^{aa'}(r) + y f^{abc} \tilde{f}^{a'b'c'} \Phi^{bb'}(r)\Phi^{cc'}(r) = 0, \tag{6.24}$$

where r is the radial coordinate. For simplicity, we may assume that both gauge groups are the same, and look for a solution of the form

$$\Phi^{aa'} = \delta^{aa'} S(r). \tag{6.25}$$

Upon substituting this into the field equations, one finds

$$\Phi^{aa'} = -\frac{2\delta^{aa'}}{y t_A r^2}. \tag{6.26}$$

This is a genuinely non-perturbative solution, in that it involves an inverse power of the coupling.[2] To interpret it further, Ref. [194]

---

[2] Our choice of language here is ambiguous, in that one could also consider a perturbation expansion in $1/y$, in which case Eq. (6.26) could be regarded as perturbative. We will follow convention, however, in assuming "perturbative" refers to an expansion about $y = 0$.

evaluated the energy, finding

$$E = \frac{128\pi}{9} \frac{\mathcal{N}}{y^2 T_A^2} \frac{1}{r_0^3}, \tag{6.27}$$

where $\mathcal{N}$ is the dimension of the (common) Lie group and r_0 is a short-distance cut-off to avoid the divergence of the field as $r \to 0$. The energy is well behaved as $r \to \infty$, and thus Eq. (6.26) represents some sort of monopole, localised at the origin. More general such solutions can be found by assuming that the gauge group is SU(2)×SU(2). The reason for this is that the structure constants of SU(2) are simply the Levi-Civita symbol ϵ^{abc}, and one may write the more general ansatz

$$\Phi^{aa'} = A(r)\delta^{aa'} + B(r)x^a x^{a'} + C(r)\epsilon^{aa'd}x^d, \tag{6.28}$$

such that there is a mixing between spacetime and colour indices, analogous to certain solutions in non-abelian gauge theory [227, 228]. Substituting Eq. (6.28) leads to a series of coupled non-linear differential equations for the various coefficient functions, and looking for a power-like solution yields [194]

$$\Phi^{aa'} = \frac{1}{yr^2}\left[-k\left(\delta^{aa'} - \frac{x^a x^{a'}}{r^2}\right) \pm \sqrt{2k - k^2}\frac{\epsilon^{aa'd}x^d}{r}\right], \tag{6.29}$$

where k is an arbitrary constant. Given the mixing between spacetime and colour indices in this solution, Ref. [194] speculated that the biadjoint monopole may be related to its nearest equivalent in pure YM theory, namely the *Wu–Yang monopole* [229]. In a gauge such that $A_0^a = 0$, the latter solution may be written as

$$A_i^a = -\frac{\epsilon_{iak}x^k}{gr^2}, \tag{6.30}$$

where $i \in \{1, 2, 3\}$ is a spatial index and g is the coupling constant. Tempting though this interpretation might be, Ref. [111] argued convincingly that it cannot be true: there is a singular gauge transformation that takes the Wu–Yang monopole to an abelian-like solution that is closely related to the *Dirac magnetic monopole* in electromagnetism. As described in Chapter 4, the double and zeroth copies of the Dirac monopole are well known, and correspond to a pure NUT charge in gravity. The zeroth copy is a solution of the linearised field equation and thus cannot correspond to the non-linear monopoles found above. Reference [200] extended these arguments to general non-abelian gauge groups.

As well as purely power-like solutions, Ref. [195] found extended monopole-like solutions, in which the divergence of the field at the origin is partially screened.[3] Moving away from spherical symmetry, Ref. [196] found biadjoint scalar solutions corresponding to extended string-like objects. Examples included those with a pure power-like dependence on the cylindrical radial coordinate, as well as dressed solutions, thus mimicking the spherically symmetric case of Refs. [194, 195].

Reference [203] attempted to apply similar methods to biadjoint scalar field theory in four-dimensional Euclidean, rather than Minkowski, spacetime. The reason for this is that there are very well-known solutions of the Euclidean field equations of both Yang–Mills theory and gravity, known as *instantons*. Interestingly, Ref. [203] found that there is no non-linear power-like spherically symmetric solution of Euclidean biadjoint scalar theory in four dimensions, which can ultimately be traced to the fact that such a solution can be identified with the zeroth copy of the so-called *Eguchi–Hanson instanton* in gravity [231–233]. This solution linearises the field equations and thus does not require a non-linear zeroth copy [124, 199]. Extended solutions exist, however, as well as non-trivial non-linear power-like solutions in other numbers of spacetime dimension. These may yet provide a simple testing ground for developing a non-perturbative double copy, but it has yet to be demonstrated that *any* non-linear solution of biadjoint theory can be placed into the scheme of Figure 1.2.

6.3   The Web of Double-Copiable Theories

In thinking about a potential non-perturbative double copy, it helps to ask quite how general the idea is, in terms of the theories that

---

[3]The remaining singularity in the solutions of Ref. [195] can be understood as a consequence of *Derrick's theorem* [230], which prohibits finite energy solutions of scalar field theories in four spacetime dimensions.

it connects. We have so far seen that the scheme of Figure 1.2 applies to different types of gauge and gravity theories. We might have additional matter content, for example, that is constrained by properties, such as supersymmetry. But this is by no means the most general type of theory that one can consider. There is an ever-increasing web of field theories that we know to be related by double-copy-like correspondences, such that the correct way to view Figure 1.2 is as part of some broader over-arching scheme. It is not yet known whether arbitrary theories can be obtained as copies of other theories. If not, then we also lack a detailed understanding of precisely which type of theories can be double-copied. Might there be some special principle that picks out "copiable" theories from others? And, if so, does this tell us that these theories are somehow to be favoured in the universe we live in? The next few years are likely to see fervent activity in this area, and here we will simply provide an introduction to the different types of theories that can be found in the double-copy literature.

6.3.1 *Actions and Lagrangians*

Throughout this section, we will assume a more complete knowledge of field theory than has been assumed in Chapter 2 and briefly state the additional concepts we require here. First, let us note that the state of a system of particles can be described by a set of independent *generalised coordinates* $\{q_i\}$ and their velocities $\{\dot{q}_i\}$, where the dot denotes differentiation with respect to time. Examples of generalised coordinates include linear positions, angles, and combinations of dimensionally equivalent coordinates. We may then define the *Lagrangian*

$$L(\{q_i\}, \{\dot{q}_i\}) = T - V, \qquad (6.31)$$

where T and V are the total kinetic and potential energies, respectively. Let the system be characterised by particular values of the (generalised) coordinates and velocities at initial and final times t_0 and t_1. Then one may define the *action*

$$S = \int_{t_0}^{t_1} dt L(\{q_i\}, \{\dot{q}_i\}), \qquad (6.32)$$

which is extremised for classical motions of the system. The latter is known as the *principle of least action* and turns out to imply the equations of motion:

$$\frac{d}{dt}\left(\frac{\partial L}{\partial \dot{q}_i}\right) = \frac{\partial L}{\partial q_i}. \tag{6.33}$$

There is one equation for each generalised coordinate, and the resulting system of coupled equations is known as the *Euler–Lagrange equations*. The principle of least action is readily generalised to field theories, where the value of a field $\phi(x)$ at a given point in spacetime can be thought of as an independent degree of freedom analogous to the generalised coordinates described earlier. However, we now have infinitely many degrees of freedom, given that there are field values at every point in spacetime. Furthermore, in a relativistically invariant theory, the equations of motion will also depend on the derivatives $\partial_\mu \phi$ in general, and the Lagrangian in this situation can be written as the volume integral of a *Lagrangian density* $\mathcal{L}$:

$$L = \int d^3x \mathcal{L}(\phi_i, \partial_\mu \phi_i), \tag{6.34}$$

where we have assumed a set of fields $\{\phi_i\}$ in general. The action is then

$$S = \int d^4x \mathcal{L}(\phi_i, \partial_\mu \phi_i), \tag{6.35}$$

and the Euler–Lagrange equations are

$$\partial_\mu \left(\frac{\partial \mathcal{L}}{\partial_\mu \phi_i}\right) = \frac{\partial \mathcal{L}}{\partial \phi_i}. \tag{6.36}$$

There is one equation for each field, such that one obtains a coupled set of non-linear equations in general. The Lagrangian also forms the basis for constructing the quantum version of the theory, to which we refer the non-expert reader to standard textbooks for more details [29–35].

6.3.2 *The non-linear sigma model*

The first theory we consider is called the *non-linear sigma model (NLSM)*. This first arose in the description of pions [234–236] but can also be considered more generally and abstractly. When a continuous symmetry in a field theory is broken, a result known as *Goldstone's theorem* states that a single massless field – or *Goldstone boson* – appears for each generator of the symmetry that is broken. One such broken symmetry in QCD arises from the fact that all the matter (quark) fields have a left- and a right-handed component, corresponding to whether the spin is pointing along or against the particle momentum. In the massless theory, one may rotate the left-handed up- and down-quark fields $(u, d)_L$ into each other and independently do the same for the right-handed fields $(u, d)_R$. These are complex 2×2 rotations, and thus there is a symmetry group $\mathrm{SU}(2) \times \mathrm{SU}(2)$. This is called *chiral symmetry* and is broken by the presence of the quark masses, which couple together the left- and right-handed quarks. However, an overall $\mathrm{SU}(2)$ symmetry survives, such that one expects a triplet of Goldstone bosons associated with the three broken generators. These are the pions, and they are not quite massless given that electromagnetic interactions mean that the original chiral symmetry is only approximate. They are thus known as *pseudo-Goldstone bosons*, but the idea clearly generalises beyond this special case.

Let us consider the Goldstone bosons associated with breaking any non-abelian group $G \times G$ to the group G, which may be written as

$$\boldsymbol{\Phi} = \Phi^a \mathbf{T}^a, \tag{6.37}$$

where $\{\mathbf{T}^a\}$ are the generators of G. Then the Lagrangian of the NLSM can be written as

$$\mathcal{L}_{\mathrm{NLSM}} = -\frac{1}{2} \mathrm{Tr} \left\{ (\partial_\mu \boldsymbol{\Phi}) \left(\mathbf{I} - \lambda^2 \boldsymbol{\Phi}^2 \right)^{-1} (\partial^\mu \boldsymbol{\Phi}) \left(\mathbf{I} - \lambda^2 \boldsymbol{\Phi}^2 \right)^{-1} \right\}, \tag{6.38}$$

where Tr denotes the trace, and λ is a coupling constant. This Lagrangian can itself be viewed as a reparametrisation (called the *Cayley parametrisation*) of the more elegant form

$$\mathcal{L}_{\mathrm{NLSM}} = \frac{1}{8\lambda^2} \mathrm{Tr} \left(\partial_\mu \mathbf{U}^\dagger \partial^\mu \mathbf{U} \right), \quad \mathbf{U} = (\mathbf{I} + \lambda \boldsymbol{\Phi})(\mathbf{I} - \lambda \boldsymbol{\Phi})^{-1}, \tag{6.39}$$

where $\mathbf{I}$ is the identity matrix in colour space. One may show that Eq. (6.39) is the leading term in the most general possible Lagrangian for Goldstone bosons for the above symmetry-breaking situation (see e.g. Ref. [29] for a textbook treatment). However, we also note that we may consider Eq. (6.38) as an interesting abstract field theory by itself, independent of its original motivation. By expanding Eq. (6.38) about small coupling λ, we may consider scattering amplitudes for the field $\boldsymbol{\Phi}$, as well as (perturbative) classical solutions.

6.3.3 *(Dirac–)Born–Infeld theory*

In the late 1930s, an alternative theory of the electromagnetic field was proposed by Born and Infeld, in order to try to circumvent the problem of the infinite self-energy of point-like charged particles, such as the electron [237–239]. The Lagrangian density for the theory can be written as

$$\mathcal{L}_{\mathrm{BI}} = \frac{1}{k}\left(\sqrt{(-1)^{d-1}\det(\eta_{\mu\nu} + kF_{\mu\nu})} - 1\right), \qquad (6.40)$$

in d spacetime dimensions, where det denotes the determinant and $F_{\mu\nu}$ is the electromagnetic field strength tensor. To understand the original motivation, note that the magnitude of the electric field of a point charge q in this theory turns out to be [237]

$$E(r) = \frac{q}{4\pi r_0^2}\frac{1}{1 + (r/r_0)^4}, \qquad (6.41)$$

where r is the radial coordinate and r_0 is a constant such that $r_0 \to 0$ as $k \to 0$. A plot of this function is shown in Figure 6.2 and compared

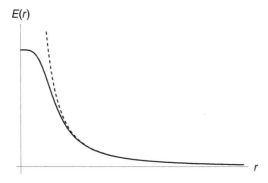

Fig. 6.2. The electric field of a point charge in the Born–Infeld theory (solid), as compared with Coulomb's law in ordinary electromagnetism (dashed).

to Coulomb's law, which is indeed recovered as the leading term in a series expansion about $r_0 = 0$. However, unlike the Coulomb result, the Born–Infeld electric field is finite as $r \to 0$. One may also show that a systematic expansion of Eq. (6.40) about $k = 0$ reproduces the known Lagrangian density for Maxwell's electromagnetism, which is then supplemented by higher-order corrections.

Born and Infeld's original motivation for this theory was to associate the electron mass with the total energy of the electromagnetic field of Eq. (6.41). However, this fixes a value of k that is incompatible with precision atomic experiments [240]. Nevertheless, interest in the Born–Infeld theory – plus its various generalisations – was rekindled by the discovery that it naturally arises from string theory [241, 242]. The latter contains extended p-dimensional objects known as *Dp-branes*, which open strings can couple to. Open strings can carry charges at their end-points, and thus the coupling of an open string with a Dp-brane leads to a gauge field on the latter. The action of this theory turns out to be a generalisation of the Born–Infeld theory, which in the case of purely bosonic string theory is described by the so-called *Dirac–Born–Infeld (DBI) action*[4]

$$\mathcal{L}_{\mathrm{DBI}} = \frac{1}{k} \left(\sqrt{(-1)^{d-1}\det(\eta_{\mu\nu} + k^2(\partial_\mu \Phi^i)(\partial_\nu \Phi^i) + kF_{\mu\nu})} - 1 \right).$$
(6.42)

If the Dp-brane has p spatial dimensions, then this Lagrangian density is to be integrated over $(p + 1)$ spacetime dimensions to yield the action. Furthermore, the $\{\Phi^i\}$ are scalar fields that correspond to transverse fluctuations of the Dp-brane. Supersymmetric generalisations of Eq. (6.42) also exist, with the resulting theory typically referred to as *Dirac–Born–Infeld–Volkov–Akulov (DBI–VA) theory* [244, 245]. Likewise, one may consider a non-abelian gauge field, rather than the abelian one considered explicitly in Eq. (6.42). Upon expanding Eq. (6.42) in the coupling k, one may consider scattering amplitudes involving the scalar and gauge field.

---

[4]We have here quoted the Lagrangian density for a Dp-brane in flat space. In a curved background, the dilaton and axion fields may also be present. See e.g. Ref. [86, 243] for textbook discussions of the DBI action.

6.3.4 (Special) Galileon theory

The phrase *Galileon theory* refers to a family of scalar field theories, in which the action is invariant under the following transformation of each scalar field:

$$\phi \to \phi + a + b \cdot x, \tag{6.43}$$

where a is a constant and b^μ is a constant 4-vector. Considering a single scalar field, Ref. [246] proved that the most general Lagrangian that gives rise to second-order equations of motion is given by

$$\mathcal{L}_{\text{Gal.}} = -\frac{1}{2}(\partial_\mu \phi)(\partial^\mu \phi) + (\partial_\mu \phi)(\partial^\mu \phi) \sum_{n=4}^{\infty} c_n \det_n, \tag{6.44}$$

where $\{c_n\}$ is a set of a arbitrary constants, and we have defined

$$\det_n = n! \partial^{[\mu_1} \partial_{\mu_1} \phi \ldots \partial^{\mu_n]} \partial_{\mu_n} \phi, \tag{6.45}$$

which can be seen to be the determinant of a certain matrix of field derivatives. Special cases of this theory first arose in the study of modifications to general relativity involving extra dimensions [247]: by considering our world as constituting a four-dimensional brane living in a five-dimensional spacetime, the gravitational dynamics associated with the fifth dimension can show up as an additional scalar field in our four-dimensional worldview. The symmetries obeyed by the underlying five-dimensional theory end up constraining the scalar field ϕ to obey the so-called *Galileon symmetry* of Eq. (6.43), and Ref. [246] then abstracted this symmetry to find its general consequences, leading to Eq. (6.44). Other work has shown that Galileon theories arise as limits of theories with massive gravitons [248] and those involving branes in a five-dimensional spacetime [249].

The term *special Galileon theory* refers to a Galileon theory with a particular choice for the coefficients $\{c_n\}$ in Eq. (6.44), such that the theory has an enhanced symmetry over and above that of Eq. (6.43). That is, the equations of motion remain invariant under field transformations of the form

$$\phi \to \phi + \theta^{\mu\nu} \Big[\Lambda x_\mu x_\nu - (\partial_\mu \phi)(\partial_\nu \phi) \Big], \tag{6.46}$$

where Λ is a constant and $\theta^{\mu\nu}$ is a constant traceless symmetric tensor. This theory first arose in Ref. [250], which used scattering amplitudes to systematically classify scalar field theories and showed that the special Galileon theory results from a certain choice for how low-energy ("soft") radiation can behave. Reference [251] found the same theory occurring naturally in a double-copy context, as we describe in the following. The presence of the enhanced symmetry of Eq. (6.46) was explained in Ref. [252] and further clarified in Ref. [253].

6.3.5 *Towards the double-copy web*

The theories of the previous few sections will look very exotic to the uninitiated. They also span many different types of physics, with entirely different historical motivations. Remarkably, however, their amplitudes can be related using very similar ideas to the double copy described in Chapter 3. Reference [251] was the first to explore this, using an approach to scattering amplitudes known as the *CHY equations* [254, 255]. A full exposition of this ideas is beyond the scope of this book, but the basic ideas are relatively simple to state. We will consider a theory of massless interacting particles and denote by $A_n[12 \ldots n]$ the (tree-level) amplitude for scattering particles with momenta $\{k_i\}$, where all colour information has been stripped off where necessary. There then exists an abstract map from the momenta $\{k_i\}$ to a set of points on a unit Riemann sphere, defined via the so-called *scattering equations* (first presented in a different context in Ref. [256]):

$$\sum_{j \neq i} \frac{k_i \cdot k_j}{\sigma_{ij}} = 0, \quad \sigma_{ij} = \sigma_i - \sigma_j.$$
(6.47)

Here, there are n equations, whose solution yields a set of n complex numbers $\{\sigma_i\}$ i.e. the locations of the points on the sphere. References [254, 255] argued that (colour-stripped) amplitudes in a variety of theories can be expressed according to the integral formula

$$A_n[12 \ldots n] = \int \frac{d\sigma_1 d\sigma_2 \ldots d\sigma_n}{\mathrm{Vol}[\mathrm{SL}(2, \mathbb{C})]} \left[\prod_{i=1}^{n} \delta \left(\sum_{j \neq i} \frac{k_i \cdot k_j}{\sigma_{ij}} \right) \right]$$

$$\times \frac{\mathcal{I}(\{k_i\}, \{\epsilon_i\}, \{\sigma_i\})}{\sigma_{12} \sigma_{23} \ldots \sigma_{n1}}.$$
(6.48)

The integral is over all possible positions of the points $\{\sigma_i\}$, where the denominator in the measure removes any overcounting due to possible reparametrisations of the sphere (n.b. the group of such reparametrisations is SL(2,$\mathbb{C}$), as discussed in Chapter 5). The delta functions impose the scattering equations of (6.47). Finally, the rest of the integrand contains a numerator that depends on the momenta, the polarisation vectors (if relevant), and the positions of the points $\{\sigma_i\}$. This depends on the theory being considered, and Refs. [254, 255] gave an explicit prescription for pure Yang–Mills theory and gravity, as well as biadjoint scalar field theory. Finally, the denominator in Eq. (6.48) contains a certain prescribed combination of differences of the $\{\sigma_i\}$ variables that is common to all theories.

The relevance of Eq. (6.48) for the double copy is that the numerators for different field theories turn out to be related by simple product rules. For example, the numerator for gravity obeys the relation

$$\mathcal{I}_G = [\mathcal{I}_{\text{YM}}]^2, \tag{6.49}$$

where $\mathcal{I}_{\text{YM}}$ is the corresponding numerator for Yang–Mills theory. This is thus an alternative way of phrasing the double copy and is referred to as the *CHY double copy* in e.g. Ref. [257]. Although the form of Eq. (6.48) looks highly mysterious, it can be derived from a certain type of string theory known as *ambitwistor string theory*, which can itself be understood as a certain limit of conventional string theory [258].

Using the CHY language, Ref. [251] derived numerator factors for use in Eq. (6.48) corresponding to the NLSM, DBI, and special Galileon theories discussed above. It also generalised Eq. (6.49) to the more general statement:

$$\mathcal{I}_X = \mathcal{I}_1 \mathcal{I}_2. \tag{6.50}$$

Here, $\mathcal{I}_{1,2}$ are the numerator factors for two (possibly different) theories and $\mathcal{I}_X$ are those associated with a third theory. Combined with the theories already discussed in Figure 1.2, this gives a set of double-copy correspondences which we summarise in Table 6.1. The upper-left entry corresponds to the canonical case of $\mathcal{N} = 0$ supergravity i.e. GR plus a dilaton and axion. Aside from this, one finds that the Born–Infeld, NLSM, and special Galileon theories are also mutually related by the double copy and to Yang–Mills theory. The role of biadjoint theory (BAS) is also interesting: given that it is a

Table 6.1. Double-copy relationships between various theories, where the left-hand column and upper row correspond to the theories entering the right-hand side of Eq. (6.50), and the individual table entries give the theory that results upon combining the two numerator factors. Here, YM, NLSM, BAS, BI, and sGal refer to the Yang–Mills theory, the non-linear sigma model, the biadjoint scalar theory, the Born–Infeld theory, and the special Galileon theory, respectively.

	YM	NLSM	BAS
YM	$\mathcal{N} = 0$ SUGRA	BI	YM
NLSM	BI	sGal	NLSM
BAS	YM	NLSM	BAS

scalar theory, its colour-stripped amplitudes are merely constants. Thus, forming a double-copy product of BAS with any other theory simply returns amplitudes of the latter theory, such that biadjoint scalar theory acts as an "identity" element in the set of double-copy operations.

Although Table 6.1 was first derived using the (perturbative) CHY approach, it has subsequently been understood using other methods, which include investigation of potential kinematic algebras underlying the various theories [13, 259–261]. Supersymmetric generalisations of Table 6.1 are known, and even then this barely scratches the surface of all the theories that are currently known to be obtainable by double-copy maps. It is even possible, for example, to obtain string theory amplitudes by using double-copy formulae involving field theory data [262, 263]. The web of theories related by double-copy relationships is continually increasing, and recent summaries at the time of writing may be found in Refs. [14, 257]. An increased understanding of precisely which field theories are amenable to being copied may itself lead to a greater non-perturbative understanding of the double copy itself.

Chapter 7

Perturbative Classical Solutions

In the previous three chapters, we have focused on classical double copies that operate on exact classical solutions, due principally to the fact that this may lead to new insights into the ultimate origin and scope of the double copy. As we have seen, though, exact classical double copies are demonstrably rare, as they demand that solutions be algebraically special. This has led to a somewhat prevalent misconception that classical double copies cannot be extended to arbitrary classical solutions in gauge theory and gravity, whereas in fact a number of methods for achieving this have been developed. The price one must pay, as for the scattering amplitudes discussed in Chapter 3, is that one must work order-by-order in perturbation theory. This is not a serious restriction for gravity solutions of astrophysical interest (e.g. for gravitational wave physics), given that physically relevant situations are sufficiently messy as to require perturbative methods even when using traditional techniques. The aim of this chapter is to summarise what is known about perturbative classical double copies and to examine one of the main applications, namely to gravitational wave physics. As a byproduct, we also understand in more detail the relationship between the Kerr–Schild and amplitude double copies. Let us begin by seeing how spacetime fields can be calculated using similar methods to the scattering amplitudes discussed in Chapter 3.

7.1   Fields from Feynman Rules

In Chapter 3, we introduced scattering amplitudes – the key quantities in quantum field theory that are directly related to probabilities of given particle interactions. We also saw that these can be obtained in a highly systematic way using Feynman rules, where the latter look different in different theories (e.g. gauge theory or gravity). A given amplitude describes a collection of initial-state particles that emerge from infinitely far away in the far past, before interacting and producing final-state particles that travel infinitely far away into the future. This is very different to what happens when one wants to evaluate the classical field generated by a given source, where the latter may consist either of some isolated particle(s) or a continuous charge or mass distribution. In that case, there is a specific point in spacetime at which one wants to know the value of the field, which is manifestly not at spatial infinity and is also at a finite time. Crucially, however, we can still represent this situation using Feynman diagrams and rules. An example is shown in Figure 7.1(a), which shows a static particle emitting a gluon of 4-momentum k. The current associated with the particle is that of Eq. (4.25), whose coupling to the gluon field is of the form

$$i\mathbf{A}_\mu \mathbf{j}^\mu.$$

Using standard QFT arguments, one may show that this leads to a Feynman rule for emission of a gluon from any point on the particle worldline, which is simply given by

$$-i\mathbf{j}^\mu. \tag{7.1}$$

We can then translate the Feynman diagram of Figure 7.1(a) into an algebraic expression for the field A_μ by combining this Feynman rule with the propagator for the emitted gluon. The latter is gauge-dependent as discussed in Chapter 2, and using the Feynman gauge result of Eq. (3.7) we find that the gluon field in momentum space takes the form

$$\tilde{\mathbf{A}}_\mu(k) = -\frac{\eta_{\mu\nu}}{k^2}\tilde{\mathbf{j}}^\nu(k), \tag{7.2}$$

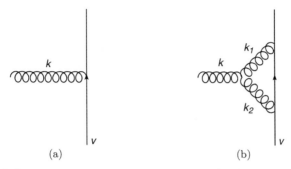

(a)   (b)

Fig. 7.1. (a) A static particle with 4-velocity $v^\mu = (1, \mathbf{0})$ emits a gluon, whose associated field $\mathbf{A}_\mu(x)$ can be evaluated at a given spacetime point x; (b) diagrams with more vertices lead to higher-order (quantum or classical) corrections to the field in perturbation theory.

where $\tilde{j}^\nu(k)$ is the Fourier transform (to momentum space) of the position-space current:

$$
\tilde{j}^\nu(k) = \int d^4x\, \mathbf{j}^\nu(x) e^{ik\cdot x}
$$

$$
= -g\, \mathbf{c}\, v^\nu \int d^4x\, \delta^{(3)}(\mathbf{x}) e^{ik\cdot x}
$$

$$
= -2\pi g\, \mathbf{c}\, v^\nu \delta(k^0). \tag{7.3}
$$

Substituting this into Eq. (7.2) and performing the inverse Fourier transform, one finds

$$
\mathbf{A}_\mu(x) = \frac{g\, \mathbf{c}\, v_\mu}{4\pi|\mathbf{x}|} = \left(\frac{g\mathbf{c}}{4\pi|\mathbf{x}|}, \mathbf{0}\right). \tag{7.4}
$$

This is indeed the field one expects from a static source, and happens to agree with the gauge choice of Eq. (4.21), such that the only non-zero component is the Coulomb form of the electrostatic potential energy. Although we have considered the simple case of a static emitting particle here, the calculation can be readily generalised to the case of an arbitrary particle world-line. In electromagnetism, this leads to the well-known *Liénard–Wiechert potential* of a moving charged particle.

In comparing the earlier analysis with the scattering amplitudes discussed in Chapter 3, we see that there are two main differences when calculating the value of a field. First, one does not include an external polarisation vector for the emitted particle. Second, in the case of scattering amplitudes, one does not include propagators for external lines, a fact which physically corresponds to the point that they propagate out to infinite distances. Here, however, we must indeed include a propagator, which indicates that the emitted particle is "off-shell", with a finite virtuality k^2. This is entirely consistent with the uncertainty principle: a gluon that does not propagate to an infinite spacetime distance has a finite length scale associated with it, which is in turn inversely related to the virtuality (i.e. an uncertainty in energy/momentum). As the gluon propagates to larger and larger distances (and hence larger times), its virtuality k^2 approaches zero, such that it goes on-shell. This is the limit encountered when calculating scattering amplitudes. Note that, unlike the latter, the presence of (gauge-dependent) propagators for external legs means that our final results will themselves be gauge-dependent. This should not worry us: we know from Chapter 2 that the fields themselves are gauge-dependent. Which gauge we obtain is fixed once we choose the convention for the Feynman rules.

The upshot of the earlier discussion is that we can think of Feynman diagrams for field values as being directly analogous to diagrams for amplitudes but where one of the legs is left off-shell, rather than all legs being on-shell. With this in mind, it immediately follows that there will be corrections to Figure 7.1(a), consisting of Feynman diagrams with higher numbers of vertices. As in Chapter 3, each vertex is associated with a power of the coupling constant (g in the earlier example), such that these additional diagrams correspond to performing a perturbation expansion. These diagrams will generate both classical and quantum corrections in general, where to fully distinguish the latter, one must include explicit factors of $\hbar$ rather than work in natural units. An example of a higher-order diagram (at $\mathcal{O}(g^3)$ in the coupling) is shown in Figure 7.1(b). Furthermore, although we have focused on Yang–Mills theory, all of our earlier remarks clearly apply in *any* (quantum) field theory. In particular, we may also carry out a similar analysis in general relativity. For the case of a single static massive particle, Ref. [264] carried out this exercise up to the first subleading order in the gravitational coupling κ, using the de Donder gauge of Eq. (5.119). The result was a perturbative construction of the Schwarzschild black hole metric, as must be the case given the uniqueness of this solution in a static, spherically symmetric scenario. This is itself interesting, given the results of Chapter 4. In Eq. (4.19), we saw that it is possible to cast the Schwarzschild solution for the metric in a form that was manifestly linear in the gravitational coupling κ, with no higher-order corrections. Upon transforming to a different coordinate system, however, the form of the metric will become non-linear in general, such that it indeed has an all-order expansion in κ. It is this expansion that is reconstructed by considering Feynman rules associated with a given coordinate choice. This in turn gives us a new point of view on how the Kerr–Schild double copy for classical solutions is related to the original BCJ double copy for scattering amplitudes. As we saw earlier, we can think of obtaining fields as an off-shell generalisation of scattering amplitude calculations, where the latter do indeed obey a double copy. If a given metric has a Kerr–Schild form, this tells us that there is a particular coordinate choice (the Kerr–Schild coordinates), such that only a single Feynman diagram becomes relevant – the gravitational analogue of Figure 7.1(a) – with all higher-order diagrams evaluating to zero. The Kerr–Schild double copy would then correspond, at least in principle, to identifying the relevant Feynman rules in the Kerr–Schild coordinate system and showing that the relevant expressions for the diagram of Figure 7.1(a) in different theories obey a double copy. It is difficult to make this argument precise, however, due to the fact that one normally fixes a coordinate choice in gravity by adding additional "gauge-fixing" terms to the gravity Lagrangian. However, it is not possible to write a general "Kerr–Schild gauge-fixing term", given that not all solutions of GR have a Kerr–Schild form. Indeed, alternative arguments linking the Kerr–Schild double copy with the BCJ double copy for scattering amplitudes have already been given in Chapter 5.

7.2 The Double Copy for Perturbative Fields

The remarks of the previous section suggest a general picture for fields in different theories. If an exact solution for a given source distribution is not possible, one may generate solutions for the fields order-by-order in perturbation theory, such that each solution may be calculated using Feynman diagrams (or some other method).

Given the similarity of this to the calculation of scattering amplitudes, it is then natural to ask whether there is a double copy for
field solutions that we may apply order-by-order in perturbation theory to generate gravitational field solutions from gauge theory ones.
This was first considered in Ref. [265], whose starting point was to
consider the exact Yang–Mills equation of Eq. (2.91) but to solve it
by using the perturbative ansatz

$$A_\mu^a = A_\mu^{(0)a} + g A_\mu^{(1)a} + g^2 A_\mu^{(2)a} + \cdots, \tag{7.5}$$

such that $A_\mu^{(n)a}$ is the contribution to the field that appears at $\mathcal{O}(g^n)$
in the coupling expansion, and where the zeroth-order term in this
approach corresponds to Eq. (7.2), obtained from the diagram of
Figure 7.1(a). Furthermore, Ref. [265] considered the explicit case of
arbitrarily moving colour charges, each associated with a worldline
$x_i^\alpha(\tau)$, where τ is the proper time. The corresponding current can be
written as

$$\mathbf{j}^{\mu a}(x) = \sum_i \int d\tau \mathbf{c}_i(\tau) v_i^\mu(\tau) \delta^{(4)}\left(x - x_i(\tau)\right), \tag{7.6}$$

where $v_i^\mu(\tau)$ is the 4-velocity (tangent vector to the worldline) at
proper time τ and $\mathbf{c}_i(\tau)$ the colour charge. The latter may change
along each worldline, due to the possibility that colour charge may
be radiated away by gluons. As argued in Ref. [265], the non-abelian
current conservation equation

$$D_\mu \mathbf{j}^\mu = 0 \tag{7.7}$$

implies the following equation for the components of the colour
vector:

$$\frac{dc^a}{d\tau} = g f^{abc} v^\mu A_\mu^b(x(\tau)) c^c(\tau). \tag{7.8}$$

The change in the 4-velocity $v^\mu(\tau)$ along each worldline is described
by the non-abelian generalisation of the Lorentz force equation of
Eq. (2.33):

$$\frac{dp_\mu}{d\tau} = m \frac{dv_\mu}{d\tau} = g c^a F_{\mu\nu}^a v^\nu. \tag{7.9}$$

If a given colour charge does not emit any gluons, its colour charge will remain constant, and it will follow a straight-line trajectory with constant 4-velocity v_i^μ. Thus, it is convenient to write the colour charge and 4-position along a given worldline as

$$\mathbf{c}_i(\tau) = \mathbf{c}_i^{(0)} + \bar{\mathbf{c}}_i(\tau), \quad x_i^\mu(\tau) = b_i^\mu + v_i^{(0)\mu}\,\tau + z^\mu(\tau), \qquad (7.10)$$

where $\mathbf{c}_i^{(0)}$ and $v_i^{(0)\mu}$ are the initial colour charge and 4-velocity, and b_i^μ the initial 4-position (relative to the spacetime origin). The quantities $\mathbf{c}_i(\tau)$ and $z^\mu(\tau)$ thus represent the deviations from these initial conditions. Substituting Eq. (7.10) into Eqs. (7.8) and (7.9) then gives, together with Eq. (2.91), a coupled system of non-linear differential equations that can be solved order-by-order in the coupling, where each order of the field in Eq. (7.5) influences the deviations in the worldline trajectory and colour at the next highest order. Reference [265] carried out this programme at the first subleading order for the emission of gluon radiation from a pair of worldlines, giving a standard Feynman diagrammatic interpretation of individual terms in the perturbation expansion. To discuss how the double copy can be applied, let us first note that Ref. [266] carried out a similar calculation to (and indeed inspired by) Ref. [265], for the specific case of radiation from a single, static colour charge. The first subleading correction to the field takes the form [266]

$$A^{(1)a\mu}(k) = \frac{i}{2(2\pi)^4}\frac{1}{k^2}f^{abc}\int d^4k_1 \int d^4k_2\,\delta(k + k_1 + k_2)$$

$$\times \left[(k - k_1)^\gamma\eta^{\mu\beta} + (k_1 - k_2)^\mu\eta^{\beta\gamma}\right.$$

$$\left. + (k_2 - k)^\beta\eta^{\gamma\mu}\right]A_\beta^{(0)b}A_\gamma^{(0)c}, \qquad (7.11)$$

where we have chosen to write this in terms of the zeroth-order field. This expression can be obtained either from solving the equations of motion perturbatively or by applying Feynman rules directly to the diagram of Figure 7.1(b), with 4-momenta as labelled in the figure (all defined to be outgoing from the 3-gluon vertex). To do this, one must combine the propagators for the three gluons with the 3-gluon vertex of Eq. (3.8). Upon integrating out the delta function, one obtains an integral over a single-loop momentum, commensurate with the Feynman rules presented in Section 3.2. Thus, we see explicitly that higher-order classical perturbative corrections can indeed be associated with Feynman diagrams, justifying the remarks of the previous section. At higher orders, one will get many different diagrams at a given order in perturbation theory, each of which will have a colour factor c_i and a kinematic numerator n_i, as is the case for scattering amplitudes.

Given expressions such as Eq. (7.11), Ref. [266] argued that one may obtain results in a gravity theory through the following multi-step procedure:

(i) One must ensure that the kinematic numerators for each contribution are BCJ-dual. This issue only arises at $\mathcal{O}(g^2)$ and above, as one needs at least two 3-gluon vertices to see Jacobi identities arising.

(ii) One must replace colour factors $\{c_i\}$ with the appropriate kinematic numerators $\{n_i\}$, where the latter include the factors of the zeroth-order gauge field $A_\beta^{(0)b}$.

(iii) One must multiply two copies of the kinematic numerator n_i of each diagram, before making the replacement

$$A_\beta^{(0)b}(p)A_{\beta'}^{(0)b'}(p) \to H_{\beta\beta'}^{(0)}(p). \tag{7.12}$$

Applying these steps to Eq. (7.13) leads to the formula

$$
\begin{aligned}
H^{(1)\mu\mu'}(k) = {} & \frac{i}{4(2\pi)^4}\frac{1}{k^2}\int d^4k_1 \int d^4k_2\, \delta(k+k_1+k_2) \\
& \times \left[(k-k_1)^\gamma \eta^{\mu\beta} + (k_1-k_2)^\mu \eta^{\beta\gamma} + (k_2-k)^\beta \eta^{\gamma\mu}\right] \\
& \times \left[(k-k_1)^{\gamma'} \eta^{\mu'\beta'} + (k_1-k_2)^{\mu'} \eta^{\beta'\gamma'} + (k_2-k)^{\beta'} \eta^{\gamma'\mu'}\right] \\
& \times H_{\beta\beta'}^{(0)}(k_1)H_{\gamma\gamma'}^{(0)}(k_2), \tag{7.13}
\end{aligned}
$$

which represents the first-order correction $H^{(1)\mu\mu'}$ to some zeroth-order field $H^{(0)\mu\mu'}$. As explained in Ref. [266], this field does not constitute a solution of pure gravity. To see why, we may recall from Chapter 3 that the double copy of pure Yang–Mills theory is not pure general relativity but instead GR coupled to two additional degrees of freedom, known as the dilaton and axion. The field whose perturbative coefficients are given by the $\{H_{\mu\nu}^{(i)}\}$ appearing in Eq. (7.13) must therefore be some composite field that contains each of these degrees of freedom. It is referred to as the *fat graviton* in Ref. [266], and one must then devise a suitable procedure for extracting the individual graviton, axion, and dilaton fields. The axion is described by an antisymmetric tensor[1] $B_{\mu\nu}(x)$, and the dilaton by a scalar field ϕ. At zeroth order, it is possible to choose a gauge (coordinate choice) in which the graviton, axion, and dilaton are simply given by the traceless symmetric, antisymmetric, and trace degrees of freedom of the fat graviton, respectively:

$$h_{\mu\nu}^{(0)} = \frac{1}{2}\left(H_{\mu\nu}^{(0)} + H_{\nu\mu}^{(0)}\right) - \frac{H^{(0)}}{4}\eta_{\mu\nu}, \quad B_{\mu\nu}^{(0)} = \frac{1}{2}\left(H_{\mu\nu}^{(0)} - H_{\nu\mu}^{(0)}\right),$$

$$\phi^{(0)} = H^{(0)}, \tag{7.14}$$

where $H^{(0)} = H^{(0)\mu}{}_\mu$. Gauge transformations will then generate a trace for the graviton field in general, so that the trace of the fat graviton must be decomposed into a genuine dilaton contribution, and the trace of the graviton. Reference [266] gave a pragmatic procedure for achieving this, based on the introduction of various projection tensors. Furthermore, Eq. (7.14) was generalised so as to provide a prescription for disentangling the physical fields from the fat graviton in the de Donder gauge, including at higher orders in perturbation theory. As an example, it was shown how special cases of the so-called *JNW solution* [267] could be obtained by double copying the gluon field sourced by a static colour charge.[2] This solution has zero axion, but a non-zero dilaton turned on in addition to the graviton. In conventional coordinates, it has line element

$$ds^2 = -\left(1 - \frac{\rho_0}{\rho}\right)^\gamma dt^2 + \left(1 - \frac{\rho_0}{\rho}\right)^{-\gamma} d\rho^2 + \left(1 - \frac{\rho_0}{\rho}\right)^{1-\gamma}\rho^2 d\Omega^2,$$

$$\tag{7.15}$$

---

[1]A general antisymmetric tensor has $d(d-1)/2$ components in d dimensions. However, not all of these are physically independent degrees of freedom, analogously to how the graviton is described by a symmetric tensor, but has only two physics degrees of freedom. In four dimensions, $B_{\mu\nu}$ has a single physical degree of freedom.

[2]The fact that the JNW solution is the most general double copy of a point charge has also been addressed in the double field theory framework that we discussed in Section 4.4.2 [268].

where t is a time coordinate, ρ a radial coordinate, $d\Omega^2$ the element of solid angle, and we have introduced the parameters

$$\rho_0 = \left(\frac{\kappa}{2}\right)^2 \frac{\sqrt{M^2 + Y^2}}{4\pi}, \quad \gamma = \frac{M}{\sqrt{M^2 + Y^2}}, \tag{7.16}$$

in terms of constants M and Y. The dilaton is given by

$$\phi = \frac{\kappa}{2}\frac{Y}{4\pi\rho_0} \log\left(1 - \frac{\rho_0}{\rho}\right), \tag{7.17}$$

and one may verify that in the limit $Y \to 0$ (with $M > 0$), the JNW solution reduces to the Schwarzschild metric, with mass M. This solution can be transformed to the de Donder gauge and compared with the result of double copying the gluon field due to a static point charge. This is carried out in Ref. [266] to first subleading order, which found that the special case $M = Y$ is naturally picked out. At linear order, however, one may choose arbitrary values of Y and M, such that any special case of the JNW solution can emerge from the double copy. This explains why the Kerr–Schild and Weyl double copies (each of which explicitly involves the linearised field equations with no higher corrections) are able to associate the pure Schwarzschild black hole with the double copy of a point charge.

Returning to the case of general worldlines, the earlier Ref. [265] found similar colour/kinematic replacement rules to those described earlier. However, the earlier remarks do not fully explain how one is to identify the kinematic numerator for a given diagram in the perturbation expansion: precisely which factors are we to include in the kinematic numerator? Put another way, how are we to identify which factors in a given diagram are *not* copied, upon making a transition to gravity? This was discussed in detail in Ref. [269], and we have seen the basic idea in both the amplitude double copy of Chapter 3 and the Kerr–Schild double copy of Chapter 4: one may carry out a counterpart to the gauge theory calculation in biadjoint scalar field theory [270]. This in turn identifies those terms in the gauge theory calculation that should not be modified upon performing the double copy to gravity. Reference [269] also outlined how the requirements of BCJ (colour-kinematics) duality could be satisfied at arbitrary orders in the perturbative expansion. Interesting extensions of the

approach of Ref. [265] include to trajectories of particles that form
bound states [271], spinning source particles [272, 273], and a gravity
theory which is itself coupled to Yang–Mills theory [274].

As in the earlier discussion, Refs. [265, 266] argued that double
copying Yang–Mills results unavoidably leads to perturbative solutions involving axions and dilatons in general, where the static point
charge maps to a special case of the JNW solution. It is natural to ask
whether one may remove the effects of the axion and dilaton and thus
generate results in pure general relativity, which is most commonly
required for astrophysical applications. Reference [275] addressed this
using an approach developed in Ref. [276] in the context of scattering
amplitudes. The basic idea is that one can introduce extra matter
degrees of freedom into the gauge theory such that, upon performing the double copy, they lead to *ghost contributions* in gravity that
act to cancel the contribution of the unwanted axion and dilaton.
A demonstration is given in Ref. [275], which considered the radiation field produced by a pair of particles. However, it is not yet
known whether this procedure generalises fully to arbitrary order in
perturbation theory, although it seems plausible.

The methods discussed in this section apply, in principle, to
arbitrary sources in gauge/gravity theory. Arguably, the most phe-
nomenologically relevant is the case of a current or energy–momentum
tensor corresponding to a pair of worldlines emitting gluons/gravitons
as appropriate. The major motivation to look at this situation comes
from the recent discovery of gravitational waves, as we explain in the
following section.

7.3   Gravitational Waves from Binary Systems

We have discussed gravitational waves throughout this book, and
they are a canonical solution of the field equations of general relativity. Until recently, however, direct evidence for these ripples in the
fabric of spacetime was missing, despite indirect evidence from the
slowdown of rotating astrophysical objects due to gravitational wave
emission [277, 278]. This situation changed dramatically in 2016, with
the direct detection of gravitational waves by the LIGO experiment:
see e.g. Refs. [279–282] for early discovery papers and Refs. [283, 284]
for a standard treatise on this subject. The coming years will see

ever more precise experiments coming online, as well as an increased understanding of how gravitational wave data can be used to understand existing or new types of structures in our universe. A typical astrophysical source for a LIGO event consists of a *binary system*, in which two colliding objects (e.g. black holes or neutron stars) approach each other and collide, before ultimately coalescing into a single object. Each such astrophysical event has three distinct stages:

(i) *The inspiral phase*: In this stage of the process, the two colliding objects orbit each other, emitting gravitational radiation in the process. The objects thus get closer as time progresses. At early times when the objects are far apart, the relevant gravitational fields are sufficiently weak that perturbation theory may be used.

(ii) *The merger*: When the objects are sufficiently close, they will merge into a single object. The gravitational fields involved are clearly very strong, such that an accurate theoretical description can only be obtained using a numerical solution of general relativity.

(iii) *The ringdown phase*: After the merger, the remaining object will wobble, before settling down to a steady state. These wobbles can be described as small perturbations of a more symmetric object, such that perturbation theory (on a curved background) may again be useful.

Whilst the dynamics that produce gravitational waves from such a source is highly non-linear, once the waves reach an Earth-based detector, the amplitude is sufficiently weak that a linearised gravity approximation may be used. Detectors consist of interferometers with very long arms, as shown in Figure 7.2(a). An incoming laser beam is split into two, where each beam travels a distance L before being reflected by a mirror. Upon returning to the beam splitter, the two beams interfere, before the combined beam is detected by a photodiode. If a gravitational wave passes through the detector, as shown in Figure 7.2(b), this leads to a fractional change

$$h = \frac{\Delta L}{L} \tag{7.18}$$

of each arm, which can in turn be related to the amplitudes of each independent polarisation state of the graviton field. This creates a

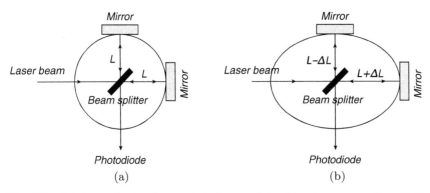

Fig. 7.2. (a) A gravitational wave detector consists of two long arms of length L. A laser beam enters and is sent by a beam splitter to two mirrors, which reflect back both beams to be ultimately measured by a photodiode; (b) a passing gravitational wave distorts the detector, such that both arms experience a fractional change in length. This creates a non-zero phase difference between the two beams, which can be measured directly.

phase difference between the two beams before they recombine, such that a non-zero interference is observed. The quantity in Eq. (7.18) is known as the *gravitational strain*, and its theoretical prediction is directly obtainable from the calculated form of the graviton field $h_{\mu\nu}$ at the detector, in a particular choice of coordinate system.

Whilst direct gravitational wave detectors measure only this quantity, other classical observables related to colliding objects are of phenomenological interest. For example, the classic Hulse–Taylor experiment of Refs. [277, 278] studied a *binary pulsar*, a type of astrophysical system in which a pulsar (a type of neutron star that emits regular strong bursts of electromagnetic radiation) orbits another object. The orbit leads to shifts in the pulsar frequency due to the Doppler effect, such that the orbital frequency of the system can be precisely determined. Continuous measurements of this system over 25 years showed that the orbital period decreased, consistent with the loss of energy due to radiation of gravitational waves. Thus, measurements of such a system are sensitive to the total power spectrum of the radiation. Other observables characterising the behaviour of binary systems include deflection angles of one of the colliding objects and/or changes in 4-momenta.

Various approaches exist for calculating gravitational wave observables in general relativity, plus extensions (see e.g.

Refs. [283, 284] for a comprehensive review). One major distinction in the literature is between methods that rely on taking a non-relativistic limit (i.e. by expanding in the velocity v of the heavy objects as a formal series in v/c) and those that do not. The former is known as the *Post-Newtonian (PN)* expansion, given that it corresponds to a series of corrections to Newtonian gravity. Keeping the full v/c dependence is referred to as the *Post-Minkowskian (PM)* expansion and can be more difficult in traditional calculations. However, as the earlier discussion has already stated, recent years have seen increased use of techniques from relativistic quantum field theory, including the application of scattering amplitudes. Such approaches naturally lead to PM results, but their validity may seem puzzling. Scattering amplitudes describe objects which collide and then separate out to infinity, which seems to be the opposite of a bound state! However, one may formally show, at least up to sub-sub-leading order in perturbation theory, that bound and unbound orbits can be formally related.[3] How this extends to higher orders in perturbation theory remains an open problem (see e.g. Ref. [285] for useful comments at the time of writing).

7.4   The Double Copy Applied to Gravitational Waves

A thorough technical introduction to (traditional) calculations for gravitational wave physics is far beyond the scope of this book, given that it would – and has – filled entire volumes by itself [283, 284]. In this section, we limit ourselves to the pervading theme of this chapter, by briefly outlining how the double copy has been used in contemporary QFT-based methods for gravitational wave physics.

7.4.1   *Worldline quantum field theory*

In Section 7.2, we have seen that the perturbative solution of the gauge or gravity equations of motion for a set of distinct particles can be expressed as a diagrammatic expansion, where certain Feynman rules arise for the emission of gluons or gravitons from

---

[3]A more precise way of saying this is that the kinematic variables of (un)bound systems can be related by an analytic continuation.

the particle worldlines. This has been formalised further in a series of papers [286–290] that introduced a *worldline quantum field theory (WQFT)* in biadjoint scalar, gauge, and gravity theories, respectively, based on the well-known Feynman path integral approach to QFT (see e.g. Refs. [29–35] for textbook treatments). A full treatment of quantum field theory is beyond the scope of this book, such that we will state the main ideas of Refs. [286–290] without full technical details. Nevertheless, the starting point for any QFT approach is to identify the appropriate classical action describing the system of interest. We saw in Chapter 6 that this provides a systematic way to derive classical equations of motion via the Euler–Lagrange equations, and there is also a well-defined procedure for taking classical actions and turning them into a full quantum theory, including the direct derivation of Feynman rules from the action itself. In each worldline QFT, the total action is given by the generic form

$$S = S_{\text{bulk}} + \sum_i S_i, \tag{7.19}$$

where S_{bulk} is the action for the biadjoint, gauge, or gravitational fields in the entirety (bulk) of the spacetime and S_i an individual action for each worldline i. The latter will depend on the trajectory $x_i^\mu(\tau)$ of each worldline, where τ is the proper time. It will also depend on the bulk fields, which may be emitted from the worldline. As an example, Ref. [290] presents a perturbative result for the worldline action for a bulk theory consisting of a graviton, axion, and dilaton:

$$S_i^{\text{grav.}} = -\frac{m_i}{2} \int d\tau \left(\dot{x}_i^2 + \kappa h_{\mu\nu} \dot{x}_i^\mu \dot{x}_i^\nu + \frac{\kappa^2}{2} h_{\mu\rho} h_\nu{}^\rho \dot{x}_i^\mu \dot{x}_i^\nu \right) + \mathcal{O}(\kappa^3), \tag{7.20}$$

where κ is the usual gravitational coupling, m_i the mass of particle i, and $x_i^\mu(\tau)$ the trajectory. The latter enters through the 4-velocity $\dot{x}_i^\mu$, where the dot denotes differentiation with respect to τ. We then see an explicit integral over the proper time, and also the presence of the graviton field $h_{\mu\nu}$, whose indices are contracted in various combinations with $\dot{x}_i^\mu$. This constitutes an explicit coupling of the graviton[4] to the worldline. For example, the term with a single power of κ in Eq. (7.20) corresponds to a single graviton emission vertex, whereas the $\mathcal{O}(\kappa^2)$ term – which has two powers of the graviton field – generates a vertex in which a correlated pair of gravitons is emitted from the same point on the worldline. Similar actions are possible in biadjoint scalar and gauge theory, where some additional complication arises due to the need to introduce colour degrees of freedom on the worldline itself.[5] Then, the total actions of Eq. (7.19) can be used to define a quantum field theory of worldlines interacting via scalar, gluon, or graviton exchange, whose Feynman diagrams consist of vertices located along each worldline, which are linked by propagators to vertices in the bulk. Example diagrams are shown in Figure 7.3. These two different types of vertices arise from the worldline and bulk actions, respectively.

Armed with these techniques, Ref. [290] shows how various observables relevant for the scattering of classical objects in gravity arise from calculating Feynman diagrams, such as those of Figure 7.3. Furthermore, algebraic expressions in the gravity case (which includes the effects of the dilaton and axion) were shown to be a double copy of a worldline QFT based on Yang–Mills theory

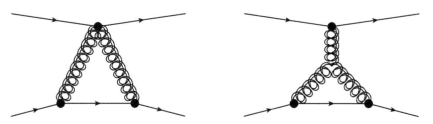

Fig. 7.3. Example Feynman diagrams generated by the worldline quantum field theory of Refs. [286–290], where we focus on the graviton case. There are vertices coupling the gravitons to the worldlines (denoted by •) and also those in the bulk of the spacetime, such as the three-graviton vertex in the second graph.

---

[4]As explained in Ref. [290], it is possible to redefine the graviton, axion, and dilaton fields such that only the graviton contributes up to $\mathcal{O}(\kappa^2)$. This is why only the graviton is present in Eq. (7.20).

[5]References [286–290] introduce fermionic fields living only on each worldline, which in turn allow one to have a scalar-valued action that nevertheless contains colour matrices. A similar approach was used in Ref. [265], discussed in the previous section, and was itself inspired by the earlier work of Ref. [291].

where, as usual, the biadjoint scalar case could be used to disentangle the relevant kinematic numerators. Reference [292] extended this approach, by showing how the WQFT could be extended to include the effects of interacting spinning particles. This in turn makes contact with traditional methods of calculation in general relativity, as follows. If the colliding objects start off by being very widely separated by some astrophysical distance, the latter can be much larger than the individual size of each object (e.g. the radius of a neutron star or the Schwarzschild radius of a black hole). It is then a good approximation to model the colliding objects as essentially point-like but where the action coupling the graviton to the worldline can take a more complicated form. To see what this looks like in general, we can think of writing the example of Eq. (7.20) as

$$S_i^{\text{grav.}} = \int d\tau \sum_j a_j \mathcal{O}_j[\dot{x}_i^\mu, h_{\mu\nu}], \qquad (7.21)$$

where each so-called *operator* $\mathcal{O}_j$ is some scalar combination of graviton fields, 4-velocities, etc., which comes with an accompanying coefficient a_i, containing coupling and numerical factors. If an extended (non-point-like) object is colliding, the trajectory $x_i^\mu(\tau)$ will describe the position of the centre of mass, where finite size effects can be modelled by including different operators $\mathcal{O}_j$ on the worldline. The latter are not arbitrary, once one applies the known constraints of (i) relativistic (Lorentz) invariance and (ii) invariance under reparametrisations of the proper time $\tau \to f(\tau)$. That is, at each order in the gravitational coupling there is only a finite set of possible operators that can be written down. A complete theory of whatever objects are colliding would in principle fix the coefficients $\{a_i\}$, such that different choices describe different types of colliding objects. However, *any* type of colliding object must, on very general grounds, have a worldline action corresponding to Eq. (7.21), in the limit in which the separation of the colliding objects is much greater than their intrinsic size. There are known operators coinciding, for example, to including spin or tidal effects. One may also introduce operators that lead to dissipative effects, where energy goes into exciting the colliding objects (e.g. surface oscillations of a neutron star). This very general framework is known as *effective field theory* (EFT) and is a well-established set of techniques for situations in which there are multiple length or energy scales, which are well separated. In gravitational scattering problems involving radiation, there is even a third length scale, namely the distance at which radiation is observed, which is much larger than either the size of the colliding objects or the distance between them. The EFT approach to gravity was first introduced in Ref. [293], and detailed reviews can be found in Refs. [294, 295]. Given the prevalence of such techniques in GR, it is natural to ask what the role of the double copy might be. One aspect – as found in Refs. [290, 292] – is that EFTs for colliding objects in gravity (at least up to some order in the operator expansion) may be obtainable as double copies of gauge theory EFTs. This in turn makes calculations in gravity EFT simpler, as one can simply recycle gauge theory results. Another role for the double copy is that, for a given theory or type of colliding object, one must calculate the explicit form of the coefficients $\{c_i\}$ appearing in Eq. (7.21). Typically, this is done by calculating scattering amplitudes (or related data) in the full theory, before taking the limit that occurs in the EFT, and then matching the two calculations to read off the values of the $\{c_i\}$. In recent work (e.g. Refs. [15–18]), such calculations have used the double copy to generate the relevant gravitational data efficiently.

7.4.2   *The KMOC formalism*

In the previous sections, we have seen how one may obtain solutions of gauge theory or gravity using a perturbative approach, where the situations we considered involved multiple colliding particles, due to the relevance for gravitational wave physics. We then want to look at certain observables characterising the collisions (e.g. the momentum of one or more of the objects, their energy, and angular momentum) and ask how this changes during the collision. Any change will be related to the fact that such systems can radiate gravitational waves, which can themselves carry away energy and (angular) momentum and in principle be observed in a detector. Reference [165] developed a systematic method for obtaining such observables directly from scattering amplitudes in quantum field theory, which has subsequently become known as the *KMOC formalism*. The starting point is to consider a given quantum state $|\psi\rangle_{\text{in}}$, representing the incoming objects. Focusing on the collision of two objects, this is typically taken to be a pair of wavepackets, separated by a spacetime displacement (or *impact parameter*) b^μ that is transverse to their incoming directions. In the far future, there will be an outgoing state

$$|\psi\rangle_{\text{out}} = \hat{S}|\psi\rangle_{\text{in}}, \tag{7.22}$$

where $\hat{S}$ is the scattering operator introduced in Eq. (3.1). As usual in quantum mechanics, the value of some observable $\mathcal{O}$ will be given by the expectation value of some associated operator $\hat{\mathcal{O}}$. The total change in the observable before and after collision will thus be a difference of expectation values, evaluated with respect to the outgoing or incoming states:

$$\Delta\mathcal{O} = {}_{\text{out}}\langle\psi|\hat{\mathcal{O}}|\psi\rangle_{\text{out}} - {}_{\text{in}}\langle\psi|\hat{\mathcal{O}}|\psi\rangle_{\text{in}}$$
$$= {}_{\text{in}}\langle\psi|\hat{S}^\dagger\hat{\mathcal{O}}\hat{S} - \hat{\mathcal{O}}|\psi\rangle_{\text{in}}. \tag{7.23}$$

To connect this with scattering amplitudes, one may introduce the decomposition of the scattering operator from Eq. (3.4), yielding

$$\Delta\mathcal{O} = {}_{\text{in}}\langle\psi|(-i\hat{T}^\dagger)\hat{\mathcal{O}} + i\hat{\mathcal{O}}\hat{T} + \hat{T}^\dagger\hat{\mathcal{O}}\hat{T}|\psi\rangle_{\text{in}}. \tag{7.24}$$

Unitarity of the scattering operator (Eq. (3.1)) implies

$$i(\hat{T} - \hat{T}^\dagger) + \hat{T}\hat{T}^\dagger = 0, \tag{7.25}$$

such that one may rewrite Eq. (7.24) as

$$\Delta\mathcal{O} = {}_{\text{in}}\langle\psi|i[\hat{\mathcal{O}}, \hat{T}]|\psi\rangle_{\text{in}} + {}_{\text{in}}\langle\psi|\hat{T}^\dagger[\hat{\mathcal{O}}, \hat{T}]|\psi\rangle_{\text{in}}. \tag{7.26}$$

This formula is an exact result but can be evaluated perturbatively using standard Feynman diagrammatic arguments, after decomposing the state $|\psi\rangle_{\text{in}}$ in terms of momentum eigenstates of the incoming objects [165]. In particular, as we saw in Eq. (3.5), matrix elements of the $\hat{T}$ operator correspond to scattering amplitudes, and we may then note the interesting property that Eq. (7.26) has a term which is *linear* in the amplitude, in addition to a term which is quadratic. This is in contrast to standard applications of scattering amplitudes in collider physics, in which measured decay rates and cross-sections are quadratic in the amplitude. Our arguments so far have been manifestly quantum. However, it is possible to systematically extract the classical limit of Eq. (7.26), as explained in detail in Ref. [165], by carefully reinstating and book-keeping factors of $\hbar$. Interestingly, in performing explicit calculations for particular observables, one often encounters spurious so-called *superclassical* contributions, containing inverse overall powers of $\hbar$ that would become singular in the classical limit $\hbar \to 0$. Such terms, however, cancel when combining the first and second terms in Eq. (7.26), although a general mechanism for this cancellation is not fully understood. Reference [165] provides examples of classical observables in electrodynamics calculated at the first subleading order in perturbation theory, namely the momentum transfer between colliding scalar particles and the radiated momentum. Reference [296] extended the formalism to describe physical quantities extracted from fields themselves. To this end, one must consider expectation values of e.g. the field strength (in gauge theory) or the Riemann curvature tensor (in a gravity theory), evaluated at the end of a collision process. Applying the earlier ideas, this yields the expectation values

$$_{\text{in}}\langle\psi|\hat{S}^\dagger \hat{F}_{\mu\nu} \hat{S}|\psi\rangle_{\text{in}}, \quad _{\text{in}}\langle\psi|\hat{S}^\dagger \hat{R}_{\mu\nu\alpha\beta} \hat{S}|\psi\rangle_{\text{in}}, \qquad (7.27)$$

where $\hat{F}_{\mu\nu}$ and $\hat{R}_{\mu\nu\alpha\beta}$ are defined in terms of the appropriate quantum field operators $\hat{A}_\mu$ (for gauge theory) and $\hat{h}_{\mu\nu}$ (for gravity), for which standard formulae exist in QFT. By again substituting the standard decomposition of $\hat{S}$, one may arrive at formulae allowing one to extract classical field values from scattering amplitudes which, as described earlier, are directly relatable to the waveforms measured by the LIGO experiment.

Regardless of the precise observable being examined, there is a clear role for the double copy in the KMOC formalism: the amplitudes entering formulae for expectation values needed for gravitational quantities can be efficiently calculated as double copies of gauge theory counterparts, using the standard BCJ double copy for scattering amplitudes.

7.4.3   *Eikonal and related methods*

There is another broad class of approaches for obtaining results pertaining to classical scattering from scattering amplitudes, each of which relies on a particular schematic form of the scattering amplitude. Let us consider again the case of two colliding objects which are separated by a transverse distance where, to be concrete, we will focus on the Yang–Mills case of two colour charges interacting via gluon exchange. If we calculate the scattering amplitude for this situation in quantum field theory at progressively higher orders in perturbation theory, we must include diagrams such as those shown in Figure 7.4, consisting of arbitrary numbers of gluon exchanges. What marks these diagrams out as special is the fact that there are multiple single-gluon exchanges between the colliding particles that may be crossed or uncrossed. Such diagrams are known as *ladder diagrams* and are distinct from diagrams, such as that in Figure 7.5, which do not have the simple ladder-like form. It is not usually physically meaningful to separate out individual types of diagrams at a given order in perturbation theory, as the result for a given diagram typically depends on the choice of gauge for the gluon field. However, something special happens in the particular kinematic limit in which the momentum transfer between the particles is much smaller than the centre of mass energy (alternatively, when the momentum carried by each exchanged gluon is much smaller than the momenta of the colliding particles). In position rather than momentum space, this

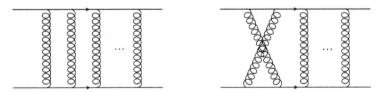

Fig. 7.4. Ladder diagrams which contribute to the interaction of two colour charges at arbitrary orders in perturbation theory.

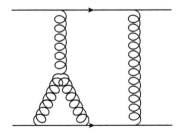

Fig. 7.5. An example of a non-ladder diagram.

coincides with the impact parameter (distance between the colliding objects) being asymptotically large, and one may then show that ladder diagrams become dominant, leading to a gauge-invariant result. Furthermore, the effect of each ladder diagram can be summed up to all orders in perturbation theory i.e. for any number of exchanged gluons, which ends up leading to the following form for the scattering amplitude in position space:

$$i\mathcal{A} = e^{i\delta} - 1. \tag{7.28}$$

Taking the limit of small momentum transfer is known as the *eikonal approximation*, and the quantity δ appearing in Eq. (7.28) is referred to as the *eikonal phase*. It is a calculable expression that can itself be expanded in the coupling constant of the theory of interest. Furthermore, the form of Eq. (7.28) can be motivated on general grounds: at any given fixed order in perturbation theory, the scattering amplitude diverges at a high energy, which violates known constraints on the high-energy behaviour of amplitudes stemming from unitarity of the quantum theory [297]. The exponential form of Eq. (7.28) – and the fact that the exponent is manifestly imaginary – means that no divergence is observed in the all-order perturbative result. Motivated by original results in non-relativistic quantum theory, this idea has a long history in quantum field theory (see e.g. Refs. [298–300] for early work in this subject). It was also applied to high-energy gravitational scattering in e.g. Refs. [301–303] and discussed in the context of generalised gravity theories (including supersymmetry) in Ref. [304], where it was firmly established that the leading behaviour of the eikonal phase in gravity is not sensitive to the additional matter content. It is known that the high-energy limits in gauge theory and gravity can be related by the double copy [78, 80–83], and thus eikonal phases in gravity can in principle be obtained from gauge theory results. It was also shown how to obtain the eikonal phase in a given theory from the worldline QFT approach of Refs. [286–290], where the gravity QFT is a manifest double copy of the gauge theory result. A known procedure exists for obtaining classical observables from the eikonal phase. For example, the deflection angle of the first incoming object in the centre of mass frame turns out to be given by

$$\theta = -\frac{1}{E_1}\frac{\partial\delta}{\partial|\boldsymbol{b}|}, \tag{7.29}$$

where E_1 is the energy of the object and $\boldsymbol{b}$ the (two-dimensional) transverse displacement between the two incoming objects. How to modify Eq. (7.28) to take into account subleading orders in perturbation theory has been a matter of some debate in the literature [18, 79, 305–313]. For example, Ref. [307] proposes an extension of the form

$$i\mathcal{A} = [1 + i\Delta] (e^{i\delta} - 1), \qquad (7.30)$$

where Δ and δ both have perturbative expansions in the coupling constant and where these are taken to collect quantum (classical) contributions, respectively. The next-to-leading result for the eikonal phase δ has been calculated for two massive objects in e.g. Refs. [79, 305, 313] and shown to reproduce known classical GR results for the deflection angle of two colliding black holes. At each higher order in perturbation theory, one must carefully disentangle quantum and classical contributions, through a rigorous accounting of factors of $\hbar$. But this is similar to how such intricacies arise in other approaches.

Given the delicacies in interpreting the eikonal phase beyond leading order, other approaches have used a different way of exponentiating the amplitude, arising from a different underlying kinematic approximation. A highly fruitful recent development at the time of writing is that of Refs. [223, 224, 314–316], which have collected terms in perturbation theory arising from the limit in which the colliding objects have asymptotically large mass, as would be appropriate for astrophysical bodies, such as black holes and neutron stars. This so-called *Heavy-mass Effective Field Theory (HEFT)* generates Feynman diagrams containing effective vertices for multiple gluon/graviton emission from each massive worldline. The scattering amplitude is argued to have a form similar to Eq. (7.28), where the eikonal phase is replaced by a so-called *HEFT phase* δ_{HEFT}. Furthermore, it is conjectured that only a restricted class of diagrams contributes to the HEFT phase, namely those which are irreducible with respect to cutting the massive worldlines (see Figure 7.6). Interestingly, the kinematic numerators of all gauge theory Feynman diagrams are found to be gauge-invariant, such that they can be made BCJ-dual i.e. manifesting the colour/kinematics duality discussed in Chapter 3. References [223, 224] have elucidated the form of the kinematic algebra that arises in this case, which itself begs the question

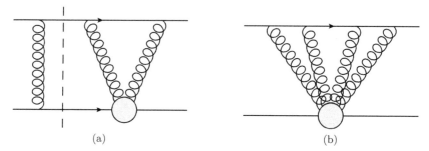

(a)   (b)

Fig. 7.6.   In the Heavy-mass Effective Field Theory (HEFT) of Refs. [223, 224, 314–316], effective vertices couple multiple gluons or gravitons to massive world-lines. The scattering amplitude for colliding objects exponentiates similarly to Eq. (7.28), where only *irreducible* diagrams contribute to the HEFT phase: (a) example of a reducible diagram, where the dashed line shows how this can be separated into two distinct subdiagrams; (b) example of an irreducible diagram.

of where this algebra ultimately comes from. It is very pleasing that a very practical question – the development of new methods for gravitational wave physics – has ended up providing a new avenue for addressing deep conceptual questions about colour–kinematics duality. As usual, once the colour–kinematics duality is manifest, one may double copy gauge theory results to a gravity theory, such that the HEFT approach provides a highly efficient way to obtain perturbative gravity results of direct astrophysical relevance.

Finally in this section, we note that Ref. [317] has introduced an alternative approach to scattering, in which the scattering operator is written in a manifestly exponential form from the outset:

$$\hat{S} = \exp\left[\frac{i}{\hbar}\hat{N}\right],\tag{7.31}$$

where $\hat{N} = \hat{N}^\dagger$ as a consequence of unitarity of $\hat{S}$. Assuming a perturbative expansion for $\hat{N}$, the latter can be related to the standard transition operator $\hat{T}$ order-by-order in the coupling so that conventional scattering amplitudes can be used to obtain a manifestly exponential representation of the S-matrix. This can then be used to obtain various classical observables, as outlined in Ref. [317].

In this chapter, we have sketched a number of ways in which perturbative classical solutions in gravity theories can be obtained as a double copy of gauge theory results. As in the cases of scattering amplitudes and exact classical solutions considered in previous chapters, the biadjoint scalar theory plays a crucial role in identifying those parts of a gauge theory result that must be replicated in order to obtain a gravity result. It is also important to stress that perturbative approaches for classical solutions are deeply related to the corresponding double copy for amplitudes (i.e. by being an off-shell generalisation) so that it should no longer be mysterious why classical double copies are possible. At the time of writing, the general picture that seems to be emerging is that double copying exact solutions is only possible in certain very special cases and that some sort of perturbative construction is to be expected for many solutions of astrophysical interest. This should not bother us unduly, given that most real-world situations require perturbation theory from the outset. The role of the double copy is then to vastly simplify the computational effort that is needed in gravity and to provide highly efficient ways of book-keeping complex calculations. The next few years are likely to see increased application of the double copy in this regard, as gravitational wave experiments become ever-more precise, necessitating higher-order calculations in perturbation theory. This itself mirrors the history of precision QCD calculations required for particle accelerator experiments, and it is the double copy that allows this knowledge and expertise to be recycled.

Chapter 8

Outlook

The beginning of the twentieth century was characterised by major problems in fundamental physics, including understanding the behaviour of light and the high-energy (short distance) behaviour of materials. The twin revolutions of quantum mechanics and relativity led to paradigm shifts in our understanding of the nature of space and time and even the definition of science itself (e.g. the role of determinism). The past century itself has seen an intense and rapid development in our understanding, where the language of relativistic quantum field theory has unified the description of fundamental forces and matter particles. Gravity is not yet included in a consistent over-arching scheme, and the initial hope and excitement from candidate ideas such as string theory and loop quantum gravity has given way to uncertainty and caution. At this point in our history, it simply does not seem to be the case that such ideas provide a straightforward and inevitable path to a "theory of everything". Despite this, string theory in particular has given us a highly intricate theoretical edifice, tying together ideas from quantum field theory in vastly different contexts. A particular idea that originated in string theory – that scattering amplitudes in gravitational theories should be related to those in (non-)abelian gauge theories [1] – underlies the subject of this book. As we have described throughout, the field theory incarnation of this idea is known as the double copy (plus related correspondences), and the past few years of highly active research have established that this idea goes way beyond its original motivation, extending to higher orders in perturbation theory and applying to classical solutions as well as potential non-perturbative information.

Furthermore, the double copy in field theory relies on an intriguing duality between colour and kinematic information in a gauge theory, which in turn implies the existence of previously hidden symmetry properties of familiar theories.

At present, we lack a full understanding of where the double copy comes from, which would in turn tell us its ultimate scope and remit. It may simply be that a suitable loop-level generalisation of the original string theory work of Ref. [1] (see e.g. Ref. [88]) provides an answer to this question. However, even if such an explanation were to be found, a number of open questions in field theory would remain: which particular (classical or quantum) quantities are expected to obey a double copy? Can double copies be obtained between arbitrary field theories and, if not, is there an underlying principle that "picks out" which field theories are relevant? Might this involve such theories being obtained as limits of string theories? Are such theories somehow favoured by nature, for this or some other reason?

The origin of the double copy is itself tied up with colour-kinematics duality, and it is certainly the case that our traditional formulation of gauge theories almost completely obscures the presence of a kinematic algebra. Indeed, it is not yet even known what the full kinematic algebra of Yang–Mills theory is, despite intriguing hints in certain special regimes. There is clearly further work to be done in pinning down kinematic algebras in particular theories, which may involve advanced mathematical ideas beyond the Lie group theory that constrains gauge theories. One possibility might be to reformulate field theories on more abstract mathematical spaces than are currently used. As an example, a well-known description of gauge theories is in terms of mathematical spaces known as *fibre bundles*. The equations of Yang–Mills theory can then be understood in terms of the known differential geometry of these structures, and their topology is also relevant for understanding different types of non-linear solutions (see e.g. Ref. [228] for a detailed review). It is natural to ask the following: is there an alternative mathematical structure, such that the kinematic algebra of Yang–Mills theory naturally arises in describing diffeomorphisms (or related transformations)? If this cannot be done for full Yang–Mills theory, can it at least be achieved for the self-dual sector? What additional insights would such a mathematical structure provide, and does string theory provide clues as to what to look for? As we reviewed in Chapter 6, there is a growing body of work phrasing colour–kinematic duality in a form that is more accessible to pure mathematicians. It will be very interesting to see how these developments evolve in the coming years, as a pure mathematical formulation of kinematic algebras has the potential to revolutionise the study of gauge theories as did the fibre bundle language in the 1970s.

The earlier remarks and open questions concern conceptual questions in field theory, and it is arguably no exaggeration to say that a full understanding of the double copy and/or colour–kinematics duality will rewrite the foundations of quantum field theory, making previously hidden structures manifest. Even if this does not happen, these ideas in their current forms have major practical applications for both gauge theory and gravity. In gauge theory, the kinematic algebra constitutes a highly non-trivial constraint that scattering amplitudes have to satisfy. Enforcing this structure may provide a shortcut when obtaining integrands for higher-order amplitudes, even in physically more relevant theories, such as QCD. As far as the double copy is concerned, by far the most pressing application of this idea in the coming years will be in generating precision calculations for gravitational wave experiments, as we reviewed in Chapter 7. This will be a major focus of global fundamental physics for the next few decades, as new and varied experiments come online. Gravitational waves provide an entirely new window through which we may view our universe. It is thus to be expected that there are new types of quantities that we will have to calculate in general relativity or alternative gravity theories, in addition to the observables we are already considering. What the double copy offers is a highly efficient way to generate intermediate steps of these calculations and to organise the ever-increasing complexity at higher orders in perturbation theory. Complex calculational tools for gauge theories have had decades of development due to successive generations of particle accelerator experiments. The fact that this vast knowledge can now be recycled to generate gravity results – whilst supporting and complementing traditional approaches – is surely a wonderful opportunity for collaboration between the fields of high energy and astrophysics. It is also extremely timely, given how recently gravitational waves have been discovered.

As well as applications in gravity, another major avenue for further work is the study of field theories in condensed matter physics.

In that context, field theories typically arise when describing effective degrees of freedom (e.g. quasiparticles) in the long wavelength limit. Some of the more exotic theories that emerge in double-copy research (e.g. biadjoint scalar field theory) may then find applications in describing certain types of materials, even if only in a restricted regime. We may also ponder whether table-top experiments providing (quantum) optical analogues of gravitons (e.g. that proposed in Ref. [318]) may be used to simulate aspects of the double copy, including axions and dilatons, or the passage of these or gravitons through media representing curved spacetime geometries. Given the present lack of understanding regarding curved-space double copies and their uses, perhaps such experiments may yet shed light on foundational aspects of the subject?

In conclusion, the double copy is a vibrant and thought-provoking research area that forces us to think about all the physics we know about in new ways. Over the past few years, an increasing variety of scientists and/or mathematicians have been seduced by the myriad charms of this subject, leading to new types of collaborations. The author's humble hope is that this book will have provided you with an accessible route to further study of the double copy, as well as the confidence to explore the many open questions at your enthusiastic leisure!

Bibliography

[1] H. Kawai, D. Lewellen and S. Tye, A relation between tree amplitudes of closed and open strings, *Nucl. Phys.* **B269**, 1 (1986). doi: 10.1016/0550-3213(86)90362-7.

[2] Z. Bern, J. Carrasco and H. Johansson, New relations for gauge-theory amplitudes, *Phys. Rev.* **D78**, 085011 (2008). doi: 10.1103/PhysRevD.78.085011, arXiv:0805.3993 [hep-ph].

[3] Z. Bern, J. J. M. Carrasco and H. Johansson, Perturbative quantum gravity as a double copy of gauge theory, *Phys. Rev. Lett.* **105**, 061602 (2010). doi: 10.1103/PhysRevLett.105.061602, arXiv:1004.0476 [hep-th].

[4] Z. Bern, T. Dennen, Y.-T. Huang and M. Kiermaier, Gravity as the square of gauge theory, *Phys. Rev.* **D82**, 065003 (2010). doi: 10.1103/PhysRevD.82.065003, arXiv:1004.0693 [hep-th].

[5] Z. Bern, S. Davies, T. Dennen and Y.-T. Huang, Absence of three-loop four-point divergences in N=4 supergravity, *Phys. Rev. Lett.* **108**, 201301 (2012). doi: 10.1103/PhysRevLett.108.201301, arXiv:1202.3423 [hep-th].

[6] Z. Bern, S. Davies, T. Dennen and Y.-T. Huang, Ultraviolet cancellations in half-maximal supergravity as a consequence of the double-copy structure, *Phys. Rev. D* **86**, 105014 (2012). doi: 10.1103/PhysRevD.86.105014, arXiv:1209.2472 [hep-th].

[7] Z. Bern, S. Davies, T. Dennen, A. V. Smirnov and V. A. Smirnov, Ultraviolet properties of n=4 supergravity at four loops, *Phys. Rev. Lett.* **111**(23), 231302 (2013). doi: 10.1103/PhysRevLett.111.231302, arXiv:1309.2498 [hep-th].

[8] Z. Bern, S. Davies and T. Dennen, The ultraviolet critical dimension of half-maximal supergravity at three loops (2014). arXiv:1412.2441 [hep-th].

[9] Z. Bern, M. Enciso, J. Parra-Martinez and M. Zeng, Manifesting enhanced cancellations in supergravity: Integrands versus integrals, *JHEP* **05**, 137 (2017). doi: 10.1007/JHEP05(2017)137, arXiv:1703.08927 [hep-th].

[10] Z. Bern, J. J. M. Carrasco, W.-M. Chen, H. Johansson, R. Roiban and M. Zeng, Five-loop four-point integrand of $N = 8$ supergravity as a generalized double copy, *Phys. Rev. D* **96**(12), 126012 (2017). doi: 10.1103/PhysRevD.96.126012, arXiv:1708.06807 [hep-th].

[11] Z. Bern, J. J. Carrasco, W.-M. Chen, A. Edison, H. Johansson, J. Parra-Martinez, R. Roiban and M. Zeng, Ultraviolet properties of \ $= 8$ supergravity at five loops, *Phys. Rev. D* **98**(8), 086021 (2018). doi: 10.1103/PhysRevD.98.086021, arXiv:1804.09311 [hep-th].

[12] J. J. M. Carrasco and L. Rodina, UV considerations on scattering amplitudes in a web of theories, *Phys. Rev. D* **100**(12), 125007 (2019). doi: 10.1103/PhysRevD.100.125007, arXiv:1908.08033 [hep-th].

[13] C. Cheung, C.-H. Shen and C. Wen, Unifying relations for scattering amplitudes, *JHEP* **02**, 095 (2018). doi: 10.1007/JHEP02(2018)095, arXiv:1705.03025 [hep-th].

[14] Z. Bern, J. J. Carrasco, M. Chiodaroli, H. Johansson and R. Roiban, The duality between color and kinematics and its applications (2019). arXiv:1909.01358 [hep-th].

[15] Z. Bern, C. Cheung, R. Roiban, C.-H. Shen, M. P. Solon and M. Zeng, Scattering amplitudes and the conservative Hamiltonian for binary systems at third post-Minkowskian order, *Phys. Rev. Lett.* **122**(20), 201603 (2019). doi: 10.1103/PhysRevLett.122.201603, arXiv:1901.04424 [hep-th].

[16] Z. Bern, J. Parra-Martinez, R. Roiban, M. S. Ruf, C.-H. Shen, M. P. Solon and M. Zeng, Scattering amplitudes and conservative binary dynamics at $\mathcal{O}(G^4)$, *Phys. Rev. Lett.* **126**(17), 171601 (2021). doi: 10.1103/PhysRevLett.126.171601, arXiv:2101.07254 [hep-th].

[17] Z. Bern, C. Cheung, R. Roiban, C.-H. Shen, M. P. Solon and M. Zeng, Black hole binary dynamics from the double copy and effective theory, *JHEP* **10**, 206 (2019). doi: 10.1007/JHEP10(2019)206, arXiv:1908.01493 [hep-th].

[18] Z. Bern, A. Luna, R. Roiban, C.-H. Shen and M. Zeng, Spinning black hole binary dynamics, scattering amplitudes, and effective field theory, *Phys. Rev. D* **104**(6), 065014 (2021). doi: 10.1103/PhysRevD.104.065014, arXiv:2005.03071 [hep-th].

[19] Z. Bern, D. Kosmopoulos, A. Luna, R. Roiban and F. Teng, Binary dynamics through the fifth power of spin at $\mathcal{O}(G^2)$ (2022). arXiv:2203.06202 [hep-th].

[20] C. D. White, Double copy—from optics to quantum gravity: Tutorial, *J. Opt. Soc. Am. B* **38**(11), 3319–3330 (2021). doi: 10.1364/JOSAB.432984, arXiv:2105.06809 [physics.optics].

[21] C. G. Böhmer, *Introduction to General Relativity and Cosmology*. World Scientific (Europe), London, UK (2016). doi: https://doi.org/ 10.1142/q0034.

[22] C. D. White, *Electromagnetism — Principles and Modern Applications*. World Scientific (Europe), London, UK (2023). doi: https:// doi.org/10.1142/q0402.

[23] C.-N. Yang and R. L. Mills, Conservation of isotopic spin and isotopic gauge invariance, *Phys. Rev.* **96**, 191–195 (1954). doi: 10.1103/PhysRev.96.191.

[24] J. Baez and J. P. Muniain, *Gauge Fields, Knots and Gravity*. World Scientific, Singapore (1995). doi: https://doi.org/10.1142/2324.

[25] D. Lovelock, The Einstein tensor and its generalizations, *J. Math. Phys.* **12**, 498–501 (1971). doi: 10.1063/1.1665613.

[26] H. Elvang and Y.-T. Huang, *Scattering Amplitudes in Gauge Theory and Gravity*. Cambridge University Press, Cambridge, UK (2015).

[27] J. M. Henn and J. C. Plefka, *Scattering Amplitudes in Gauge Theories*, Vol. 883. Springer, Berlin (2014). doi: 10.1007/ 978-3-642-54022-6.

[28] G. Travaglini *et al.*, The SAGEX review on scattering amplitudes (2022). arXiv:2203.13011 [hep-th].

[29] M. D. Schwartz, *Quantum Field Theory and the Standard Model*. Cambridge University Press, Cambridge, UK (2014).

[30] L. H. Ryder, *Quantum Field Theory*. Cambridge University Press, Cambridge, UK (1996).

[31] S. Weinberg, *The Quantum Theory of Fields. Vol. 1: Foundations*. Cambridge University Press, Cambridge, UK (2005).

[32] S. Weinberg, *The Quantum Theory of Fields. Vol. 2: Modern Applications*. Cambridge University Press, Cambridge, UK (2013).

[33] A. Zee, *Quantum Field Theory in a Nutshell*. Princeton University Press, Princeton, NJ (2003).

[34] G. F. Sterman, *An Introduction to Quantum Field Theory*. Cambridge University Press, Cambridge, UK (1993).

[35] M. Srednicki, *Quantum Field Theory*. Cambridge University Press, Cambridge, UK (2007).

[36] L. D. Faddeev and V. N. Popov, Feynman diagrams for the Yang–Mills field, *Phys. Lett.* **25B**, 29–30 (1967). doi: 10.1016/ 0370-2693(67)90067-6.

[37] L. Borsten, H. Kim, B. Jurco, T. Macrelli, C. Saemann and M. Wolf, Double copy from homotopy algebras (2021). arXiv:2102.11390 [hep-th].

[38] L. Borsten, B. Jurco, H. Kim, T. Macrelli, C. Saemann and M. Wolf, Tree-level color-kinematics duality implies loop-level color-kinematics duality (2021). arXiv:2108.03030 [hep-th].

[39] N. Bjerrum-Bohr, P. H. Damgaard and P. Vanhove, Minimal basis for gauge theory amplitudes, *Phys. Rev. Lett.* **103**, 161602 (2009). doi: 10.1103/PhysRevLett.103.161602, arXiv:0907.1425 [hep-th].

[40] S. Stieberger, Open and closed vs. pure open string disk amplitudes (2009). arXiv:0907.2211 [hep-th].

[41] N. Bjerrum-Bohr, P. H. Damgaard, T. Sondergaard and P. Vanhove, Monodromy and Jacobi-like relations for color-ordered amplitudes, *JHEP* **1006**, 003 (2010). doi: 10.1007/JHEP06(2010)003, arXiv:1003.2403 [hep-th].

[42] B. Feng, R. Huang and Y. Jia, Gauge amplitude identities by on-shell recursion relation in S-matrix program, *Phys. Lett.* **B695**, 350–353 (2011). doi: 10.1016/j.physletb.2010.11.011, arXiv:1004.3417 [hep-th].

[43] S. Henry Tye and Y. Zhang, Dual identities inside the gluon and the graviton scattering amplitudes, *JHEP* **1006**, 071 (2010). doi: 10.1007/JHEP06(2010)071,10.1007/JHEP04(2011)114, arXiv:1003.1732 [hep-th].

[44] C. R. Mafra, O. Schlotterer and S. Stieberger, Explicit BCJ numerators from pure spinors, *JHEP* **1107**, 092 (2011). doi: 10.1007/JHEP07(2011)092, arXiv:1104.5224 [hep-th].

[45] R. Monteiro and D. O'Connell, The kinematic algebra from the self-dual sector, *JHEP* **1107**, 007 (2011). doi: 10.1007/JHEP07(2011)007, arXiv:1105.2565 [hep-th].

[46] N. Bjerrum-Bohr, P. H. Damgaard, R. Monteiro and D. O'Connell, Algebras for amplitudes, *JHEP* **1206**, 061 (2012). doi: 10.1007/JHEP06(2012)061, arXiv:1203.0944 [hep-th].

[47] Z. Bern, L. J. Dixon, D. Dunbar, M. Perelstein and J. Rozowsky, On the relationship between Yang–Mills theory and gravity and its implication for ultraviolet divergences, *Nucl. Phys.* **B530**, 401–456 (1998). doi: 10.1016/S0550-3213(98)00420-9, arXiv:hep-th/9802162 [hep-th].

[48] M. B. Green, J. H. Schwarz and L. Brink, N=4 Yang–Mills and N=8 supergravity as limits of string theories, *Nucl. Phys.* **B198**, 474–492 (1982). doi: 10.1016/0550-3213(82)90336-4.

[49] Z. Bern, J. Rozowsky and B. Yan, Two loop four gluon amplitudes in N=4 superYang–Mills, *Phys. Lett.* **B401**, 273–282 (1997). doi: 10.1016/S0370-2693(97)00413-9, arXiv:hep-ph/9702424 [hep-ph].

[50] J. J. Carrasco and H. Johansson, Five-point amplitudes in N=4 super-Yang–Mills theory and N=8 supergravity, *Phys. Rev.* **D85**, 025006 (2012). doi: 10.1103/PhysRevD.85.025006, arXiv:1106.4711 [hep-th].

[51] J. J. M. Carrasco, M. Chiodaroli, M. Günaydin and R. Roiban, One-loop four-point amplitudes in pure and matter-coupled N=4 supergravity, *JHEP* **1303**, 056 (2013). doi: 10.1007/JHEP03(2013)056, arXiv:1212.1146 [hep-th].

[52] C. R. Mafra and O. Schlotterer, The structure of n-point one-loop open superstring amplitudes, *JHEP* **1408**, 099 (2014). doi: 10.1007/JHEP08(2014)099, arXiv:1203.6215 [hep-th].

[53] R. H. Boels, R. S. Isermann, R. Monteiro and D. O'Connell, Colour-kinematics duality for one-loop rational amplitudes, *JHEP* **1304**, 107 (2013). doi: 10.1007/JHEP04(2013)107, arXiv:1301.4165 [hep-th].

[54] N. E. J. Bjerrum-Bohr, T. Dennen, R. Monteiro and D. O'Connell, Integrand oxidation and one-loop colour-dual numerators in N=4 gauge theory, *JHEP* **1307**, 092 (2013). doi: 10.1007/JHEP07(2013)092, arXiv:1303.2913 [hep-th].

[55] Z. Bern, S. Davies, T. Dennen, Y.-T. Huang and J. Nohle, Color-kinematics duality for Pure Yang–Mills and gravity at one and two loops (2013). arXiv:1303.6605 [hep-th].

[56] Z. Bern, S. Davies and T. Dennen, The ultraviolet structure of half-maximal supergravity with matter multiplets at two and three loops, *Phys. Rev.* **D88**, 065007 (2013). doi: 10.1103/PhysRevD.88.065007, arXiv:1305.4876 [hep-th].

[57] J. Nohle, Color-kinematics duality in one-loop four-gluon amplitudes with matter (2013). arXiv:1309.7416 [hep-th].

[58] S. G. Naculich, H. Nastase and H. J. Schnitzer, All-loop infrared-divergent behavior of most-subleading-color gauge-theory amplitudes, *JHEP* **1304**, 114 (2013). doi: 10.1007/JHEP04(2013)114, arXiv:1301.2234 [hep-th].

[59] Y.-J. Du, B. Feng and C.-H. Fu, Dual-color decompositions at one-loop level in Yang–Mills theory (2014). arXiv:1402.6805 [hep-th].

[60] C. R. Mafra and O. Schlotterer, Towards one-loop SYM amplitudes from the pure spinor BRST cohomology, *Fortsch. Phys.* **63**(2), 105–131 (2015). doi: 10.1002/prop.201400076, arXiv:1410.0668 [hep-th].

[61] Z. Bern, S. Davies and T. Dennen, Enhanced ultraviolet cancellations in N = 5 supergravity at four loop (2014). arXiv:1409.3089 [hep-th].

[62] C. R. Mafra and O. Schlotterer, Two-loop five-point amplitudes of super Yang–Mills and supergravity in pure spinor superspace (2015). arXiv:1505.02746 [hep-th].

[63] S. He, R. Monteiro and O. Schlotterer, String-inspired BCJ numerators for one-loop MHV amplitudes, *JHEP* **01**, 171 (2016). doi: 10.1007/JHEP01(2016)171, arXiv:1507.06288 [hep-th].

[64] Z. Bern, S. Davies and J. Nohle, Double-copy constructions and unitarity cuts (2015). arXiv:1510.03448 [hep-th].

[65] G. Mogull and D. O'Connell, Overcoming obstacles to colour-kinematics duality at two loops, *JHEP* **12**, 135 (2015). doi: 10.1007/JHEP12(2015)135, arXiv:1511.06652 [hep-th].

[66] M. Chiodaroli, M. Gunaydin, H. Johansson and R. Roiban, Spontaneously broken Yang–Mills–Einstein supergravities as double copies (2015). arXiv:1511.01740 [hep-th].

[67] R. P. Feynman, Quantum theory of gravitation, *Acta Phys. Polon.* **24**, 697–722 (1963).

[68] R. P. Feynman, *Feynman Lectures on Gravitation*. Westview Press, Boulder, CO (1996).

[69] R. Arnowitt and S. Deser, Quantum theory of gravitation: General formulation and linearized theory, *Phys. Rev.* **113**, 745–750 (1959). doi: 10.1103/PhysRev.113.745.

[70] B. S. DeWitt, Quantum theory of gravity. 1. The canonical theory, *Phys. Rev.* **160**, 1113–1148 (1967). doi: 10.1103/PhysRev.160.1113.

[71] B. S. DeWitt, Quantum theory of gravity. 2. The manifestly covariant theory, *Phys. Rev.* **162**, 1195–1239 (1967). doi: 10.1103/PhysRev.162.1195.

[72] B. S. DeWitt, Quantum theory of gravity. 3. Applications of the covariant theory, *Phys. Rev.* **162**, 1239–1256 (1967). doi: 10.1103/PhysRev.162.1239.

[73] S. Mandelstam, Quantization of the gravitational field, *Ann. Phys.* **19**, 25–66 (1962). doi: 10.1016/0003-4916(62)90233-6.

[74] S. Mandelstam, Feynman rules for the gravitational field from the coordinate independent field theoretic formalism, *Phys. Rev.* **175**, 1604–1623 (1968). doi: 10.1103/PhysRev.175.1604.

[75] M. J. G. Veltman, Quantum theory of gravitation, *Conf. Proc. C* **7507281**, 265–327 (1975).

[76] S. Oxburgh and C. White, BCJ duality and the double copy in the soft limit, *JHEP* **1302**, 127 (2013). doi: 10.1007/JHEP02(2013)127, arXiv:1210.1110 [hep-th].

[77] C. D. White, Factorization properties of soft graviton amplitudes, *JHEP* **1105**, 060 (2011). doi: 10.1007/JHEP05(2011)060, arXiv:1103.2981 [hep-th].

[78] S. Melville, S. Naculich, H. Schnitzer and C. White, Wilson line approach to gravity in the high energy limit, *Phys. Rev.* **D89**, 025009 (2014). doi: 10.1103/PhysRevD.89.025009, arXiv:1306.6019 [hep-th].

[79] A. Luna, S. Melville, S. G. Naculich and C. D. White, Next-to-soft corrections to high energy scattering in QCD and gravity, *JHEP*

01, 052 (2017). doi: 10.1007/JHEP01(2017)052, arXiv:1611.02172 [hep-th].

[80] R. Saotome and R. Akhoury, Relationship between gravity and gauge scattering in the high energy limit, *JHEP* **1301**, 123 (2013). doi: 10.1007/JHEP01(2013)123, arXiv:1210.8111 [hep-th].

[81] A. Sabio Vera, E. Serna Campillo and M. A. Vazquez-Mozo, Color-kinematics duality and the Regge limit of inelastic amplitudes, *JHEP* **1304**, 086 (2013). doi: 10.1007/JHEP04(2013)086, arXiv:1212.5103 [hep-th].

[82] H. Johansson, A. Sabio Vera, E. Serna Campillo and M. A. Vazquez-Mozo, Color-kinematics duality in multi-Regge kinematics and dimensional reduction, *JHEP* **1310**, 215 (2013). doi: 10.1007/JHEP10(2013)215, arXiv:1307.3106 [hep-th].

[83] H. Johansson, A. Sabio Vera, E. Serna Campillo and M. A. Vazquez-Mozo, Color-kinematics duality and dimensional reduction for graviton emission in Regge limit (2013). arXiv:1310.1680 [hep-th].

[84] M. B. Green, J. H. Schwarz and E. Witten, *Superstring Theory. Vol. 1: Introduction*. Cambridge Monographs on Mathematical Physics. Cambridge University Press, Cambridge, UK (1988).

[85] M. B. Green, J. H. Schwarz and E. Witten, *Superstring Theory. Vol. 2: Loop Amplitudes, Anomalies and Phenomenology*. Cambridge University Press, Cambridge, UK (1988).

[86] J. Polchinski, *String Theory. Vol. 1: An Introduction to the Bosonic String*. Cambridge Monographs on Mathematical Physics. Cambridge University Press, Cambridge, UK (2007). doi: 10.1017/CBO9780511816079.

[87] J. Polchinski, *String Theory. Vol. 2: Superstring Theory and Beyond*. Cambridge Monographs on Mathematical Physics. Cambridge University Press, Cambridge, UK (2007). doi: 10.1017/CBO9780511618123.

[88] S. Stieberger, A relation between one-loop amplitudes of closed and open strings (one-loop KLT relation) (2022). arXiv:2212.06816 [hep-th].

[89] R. Monteiro, D. O'Connell and C. D. White, Black holes and the double copy, *JHEP* **1412**, 056 (2014). doi: 10.1007/JHEP12(2014)056, arXiv:1410.0239 [hep-th].

[90] V. E. Didenko, A. S. Matveev and M. A. Vasiliev, Unfolded description of AdS(4) Kerr black hole, *Phys. Lett. B* **665**, 284–293 (2008). doi: 10.1016/j.physletb.2008.05.067, arXiv:0801.2213 [gr-qc].

[91] V. E. Didenko and M. A. Vasiliev, Static BPS black hole in 4d higher-spin gauge theory, *Phys. Lett. B* **682**, 305–315 (2009). doi: 10.1016/j.physletb.2009.11.023, arXiv:0906.3898 [hep-th]. [Erratum: *Phys. Lett. B* **722**, 389 (2013)].

[92]   H. Stephani, D. Kramer, M. A. H. MacCallum, C. Hoenselaers and
       E. Herlt, *Exact Solutions of Einstein's Field Equations*. Cambridge
       Monographs on Mathematical Physics. Cambridge University Press,
       Cambridge (2003). doi: 10.1017/CBO9780511535185.
[93]   J. B. Griffiths and J. Podolsky, *Exact Space-Times in Ein-
       stein's General Relativity*. Cambridge Monographs on Mathemat-
       ical Physics. Cambridge University Press, Cambridge (2009). doi:
       10.1017/CBO9780511635397.
[94]   R. P. Kerr and A. Schild, A new class of vacuum solutions of the
       Einstein field equations. In G. Barbera (ed.) *Atti del convegno sulla
       relatività generale; problemi dell'energia e onde gravitazionali*, p. 173
       (1965).
[95]   R. P. Kerr and A. Schild, Some algebraically degenerate solutions
       of Einstein's gravitational field equations. In R. Finn (ed.) *Proceed-
       ings of Symposia in Applied Mathematics*, Vol. XVII, pp. 199–209
       (1965).
[96]   W. Israel, Source of the Kerr metric, *Phys. Rev.* **D2**, 641–646 (1970).
       doi: 10.1103/PhysRevD.2.641.
[97]   H. Balasin and H. Nachbagauer, Distributional energy momen-
       tum tensor of the Kerr–Newman space-time family, *Class. Quant.
       Grav.* **11**, 1453–1462 (1994). doi: 10.1088/0264-9381/11/6/010,
       arXiv:gr-qc/9312028 [gr-qc].
[98]   A. K. Ridgway and M. B. Wise, Static spherically symmetric Kerr–
       Schild metrics and implications for the classical double copy, *Phys.
       Rev.* **D94**(4), 044023 (2016). doi: 10.1103/PhysRevD.94.044023,
       arXiv:1512.02243 [hep-th].
[99]   R. C. Myers and M. Perry, Black holes in higher dimensional space-
       times, *Ann. Phys.* **172**, 304 (1986). doi: 10.1016/0003-4916(86)
       90186-7.
[100] J. M. Maldacena, The large N limit of superconformal field theories
       and supergravity, *Adv. Theor. Math. Phys.* **2**, 231–252 (1998). doi:
       10.1023/A:1026654312961, arXiv:hep-th/9711200.
[101] A. H. Taub, Empty space-times admitting a three parameter group
       of motions, *Ann. Math.* **53**(3), 472–490 (1951). http://www.jstor.
       org/stable/1969567.
[102] E. Newman, L. Tamburino and T. Unti, Empty-space generalization
       of the schwarzschild metric, *J. Math. Phys.* **4**(7), 915–923 (1963).
       doi: http://dx.doi.org/10.1063/1.1704018, http://scitation.aip.org/
       content/aip/journal/jmp/4/7/10.1063/1.1704018.
[103] J. F. Plebanski, A class of solutions of Einstein-Maxwell equations,
       Ann. Phys. **90**(1), 196–255 (1975). doi: http://dx.doi.org/10.1016/
       0003-4916(75)90145-1, http://www.sciencedirect.com/science/artic
       le/pii/0003491675901451.

[104] Z. Chong, G. Gibbons, H. Lu and C. Pope, Separability and killing tensors in Kerr–Taub–NUT–de sitter metrics in higher dimensions, *Phys. Lett.* **B609**, 124–132 (2005). doi: 10.1016/j.physletb.2004.07. 066, arXiv:hep-th/0405061 [hep-th].

[105] A. Luna, R. Monteiro, D. O'Connell and C. D. White, The classical double copy for Taub-NUT spacetime, *Phys. Lett.* **B750**, 272–277 (2015). doi: 10.1016/j.physletb.2015.09.021, arXiv:1507.01869 [hep-th].

[106] T. Ortin, *Gravity and Strings*. Cambridge Monographs on Mathematical Physics. Cambridge University Press, Cambridge, UK (2015). doi: 10.1017/CBO9781139019750, http://www.cambridge. org/mw/academic/subjects/physics/theoretical-physics-and-math ematical-physics/gravity-and-strings-2nd-edition.

[107] W. Chen and H. Lu, Kerr–Schild structure and harmonic 2-forms on (A)dS–Kerr–NUT metrics, *Phys. Lett.* **B658**, 158–163 (2008). doi: 10.1016/j.physletb.2007.09.066, arXiv:0705.4471 [hep-th].

[108] S. R. Coleman, Nonabelian plane waves, *Phys. Lett.* **B70**, 59 (1977). doi: 10.1016/0370-2693(77)90344-6.

[109] W. Siegel, Fields (1999). arXiv:hep-th/9912205 [hep-th].

[110] P. Aichelburg and R. Sexl, On the gravitational field of a massless particle, *Gen. Rel. Grav.* **2**, 303–312 (1971). doi: 10.1007/ BF00758149.

[111] N. Bahjat-Abbas, R. Stark-Muchão and C. D. White, Monopoles, shockwaves and the classical double copy (2020). arXiv:2001.09918 [hep-th].

[112] M. Carrillo-González, R. Penco and M. Trodden, The classical double copy in maximally symmetric spacetimes, *JHEP* **04**, 028 (2018). doi: 10.1007/JHEP04(2018)028, arXiv:1711.01296 [hep-th].

[113] A. Luna, R. Monteiro, I. Nicholson, D. O'Connell and C. D. White, The double copy: Bremsstrahlung and accelerating black holes, *JHEP* **06**, 023 (2016). doi: 10.1007/JHEP06(2016)023, http://arxiv. org/abs/1603.05737arxiv.org/abs/1603.05737 [hep-th].

[114] G. Aldazabal, D. Marques and C. Nunez, Double field theory: A pedagogical review, *Class. Quant. Grav.* **30**, 163001 (2013). doi: 10.1088/0264-9381/30/16/163001, arXiv:1305.1907 [hep-th].

[115] K. Lee, Kerr–Schild double field theory and classical double copy, *JHEP* **10**, 027 (2018). doi: 10.1007/JHEP10(2018)027, arXiv:1807.08443 [hep-th].

[116] W. Cho and K. Lee, Heterotic Kerr–Schild double field theory and classical double copy, *JHEP* **07**, 030 (2019). doi: 10.1007/ JHEP07(2019)030, arXiv:1904.11650 [hep-th].

[117]   K. Cho, K. Kim and K. Lee, The off-shell recursion for gravity and the classical double copy for currents, *JHEP* **01**, 186 (2022). doi: 10.1007/JHEP01(2022)186, arXiv:2109.06392 [hep-th].

[118]   T. Adamo, E. Casali, L. Mason and S. Nekovar, Scattering on plane waves and the double copy, *Class. Quant. Grav.* **35**(1), 015004 (2018). doi: 10.1088/1361-6382/aa9961, arXiv:1706.08925 [hep-th].

[119]   T. Adamo, E. Casali, L. Mason and S. Nekovar, Amplitudes on plane waves from ambitwistor strings (2017). arXiv:1708.09249 [hep-th].

[120]   T. Adamo, E. Casali and S. Nekovar, Ambitwistor string vertex operators on curved backgrounds, *JHEP* **01**, 213 (2019). doi: 10.1007/JHEP01(2019)213, arXiv:1809.04489 [hep-th].

[121]   T. Adamo, E. Casali, L. Mason and S. Nekovar, Plane wave backgrounds and colour-kinematics duality, *JHEP* **02**, 198 (2019). doi: 10.1007/JHEP02(2019)198, arXiv:1810.05115 [hep-th].

[122]   T. Adamo and A. Ilderton, Classical and quantum double copy of back-reaction, *JHEP* **09**, 200 (2020). doi: 10.1007/JHEP09(2020) 200, arXiv:2005.05807 [hep-th].

[123]   N. Bahjat-Abbas, A. Luna and C. D. White, The Kerr–Schild double copy in curved spacetime, *JHEP* **12**, 004 (2017). doi: 10.1007/JHEP12(2017)004, arXiv:1710.01953 [hep-th].

[124]   A. Luna, R. Monteiro, I. Nicholson and D. O'Connell, Type D spacetimes and the Weyl double copy, *Class. Quant. Grav.* **36**, 065003 (2019). doi: 10.1088/1361-6382/ab03e6, arXiv:1810.08183 [hep-th].

[125]   R. Penrose and W. Rindler, *Spinors and Space-Time*. Cambridge Monographs on Mathematical Physics. Cambridge University Press, Cambridge (2011). doi: 10.1017/CBO9780511564048.

[126]   J. Stewart, *Advanced General Relativity*. Cambridge Monographs on Mathematical Physics. Cambridge University Press, Cambridge, UK (1994). doi: 10.1017/CBO9780511608179.

[127]   L. D. Landau and E. M. Lifschits, *The Classical Theory of Fields. Volume 2: Course of Theoretical Physics*. Pergamon Press, Oxford (1975).

[128]   H. Stephani and J. Stewart, *General Relativity. An Introduction to the Theory of the Gravitational Field* (1982).

[129]   M. Walker and R. Penrose, On quadratic first integrals of the geodesic equations for type [22] spacetimes, *Commun. Math. Phys.* **18**, 265–274 (1970). doi: 10.1007/BF01649445.

[130]   L. P. Hughston, R. Penrose, P. Sommers and M. Walker, On a quadratic first integral for the charged particle orbits in the charged Kerr solution, *Commun. Math. Phys.* **27**, 303–308 (1972). doi: 10.1007/BF01645517.

[131]   W. Dietz and R. Rüdiger, Space-times admitting Killing-Yano tensors, *Proc. R. Soc. Lond. A* **375**, 361–378 (1981). doi: 10.1098/rspa. 1981.0056.

[132]   J. F. Plebanski and M. Demianski, Rotating, charged, and uniformly accelerating mass in general relativity, *Ann. Phys.* **98**, 98–127 (1976). doi: 10.1016/0003-4916(76)90240-2.

[133]   H. Godazgar, M. Godazgar, R. Monteiro, D. Peinador Veiga and C. N. Pope, Weyl double copy for gravitational waves, *Phys. Rev. Lett.* **126**(10), 101103 (2021). doi: 10.1103/PhysRevLett.126.101103, arXiv:2010.02925 [hep-th].

[134]   S. Sabharwal and J. W. Dalhuisen, Anti-self-dual spacetimes, gravitational instantons and knotted zeros of the Weyl tensor, *JHEP* **07**, 004 (2019). doi: 10.1007/JHEP07(2019)004, arXiv:1904.06030 [hep-th].

[135]   R. Alawadhi, D. S. Berman, B. Spence and D. Peinador Veiga, S-duality and the double copy, *JHEP* **03**, 059 (2020). doi: 10.1007/ JHEP03(2020)059, arXiv:1911.06797 [hep-th].

[136]   R. Alawadhi, D. S. Berman and B. Spence, Weyl doubling, *JHEP* **09**, 127 (2020). doi: 10.1007/JHEP09(2020)127, arXiv:2007.03264 [hep-th].

[137]   G. Elor, K. Farnsworth, M. L. Graesser and G. Herczeg, The Newman-Penrose map and the classical double copy, *JHEP* **12**, 121 (2020). doi: 10.1007/JHEP12(2020)121, arXiv:2006.08630 [hep-th].

[138]   C. D. White, Twistorial foundation for the classical double copy, *Phys. Rev. Lett.* **126**(6), 061602 (2021). doi: 10.1103/PhysRevLett. 126.061602, arXiv:2012.02479 [hep-th].

[139]   E. Chacón, S. Nagy and C. D. White, The Weyl double copy from twistor space, *JHEP* **05**, 2239 (2021). doi: 10.1007/JHEP05(2021) 239, arXiv:2103.16441 [hep-th].

[140]   E. Chacón, S. Nagy and C. D. White, Alternative formulations of the twistor double copy, *JHEP* **03**, 180 (2022). doi: 10.1007/ JHEP03(2022)180, arXiv:2112.06764 [hep-th].

[141]   E. Chacón, A. Luna and C. D. White, Double copy of the multipole expansion, *Phys. Rev. D* **106**(8), 086020 (2022). doi: 10.1103/ PhysRevD.106.086020, arXiv:2108.07702 [hep-th].

[142]   K. Farnsworth, M. L. Graesser and G. Herczeg, Twistor space origins of the Newman-Penrose map (2021). arXiv:2104.09525 [hep-th].

[143]   R. Monteiro, D. O'Connell, D. Peinador Veiga and M. Sergola, Classical solutions and their double copy in split signature, *JHEP* **05**, 268 (2021). doi: 10.1007/JHEP05(2021)268, arXiv:2012.11190 [hep-th].

[144] R. Monteiro, S. Nagy, D. O'Connell, D. Peinador Veiga and M. Sergola, NS-NS spacetimes from amplitudes, *JHEP* **06**, 021 (2022). doi: 10.1007/JHEP06(2022)021, arXiv:2112.08336 [hep-th].

[145] R. Penrose, Twistor algebra, *J. Math. Phys.* **8**, 345 (1967). doi: 10.1063/1.1705200.

[146] R. Penrose and M. A. H. MacCallum, Twistor theory: An approach to the quantization of fields and space-time, *Phys. Rep.* **6**, 241–316 (1972). doi: 10.1016/0370-1573(73)90008-2.

[147] R. Penrose, Twistor quantization and curved space-time, *Int. J. Theor. Phys.* **1**, 61–99 (1968). doi: 10.1007/BF00668831.

[148] R. Penrose and W. Rindler, *Spinors and Space-Time. Vol. 2: Spinor and Twistor Methods in Space-Time Geometry.* Cambridge Monographs on Mathematical Physics. Cambridge University Press, Cambridge, UK (1988). doi: 10.1017/CBO9780511524486.

[149] S. Huggett and K. Tod, *An Introduction to Twistor Theory.* Cambridge University Press, Cambridge, UK (1986).

[150] T. Adamo, Lectures on twistor theory, *PoS* **Modave2017**, 003 (2018). doi: 10.22323/1.323.0003, arXiv:1712.02196 [hep-th].

[151] C. Nash and S. Sen, *Topology and Geometry for Physicists.* Dover Publications, New York (1988). http://www.amazon.com/ Topology-Geometry-Physicists-Charles-Nash/dp/0125140819/ref= sr_1_1?ie=UTF8&s=book&qid=1263990064&sr=1-1.

[152] E. Witten, Perturbative gauge theory as a string theory in twistor space, *Commun. Math. Phys.* **252**, 189–258 (2004). doi: 10.1007/ s00220-004-1187-3, arXiv:hep-th/0312171.

[153] M. G. Eastwood, R. Penrose and R. O. Wells, Cohomology and massless fields, *Commun. Math. Phys.* **78**, 305–351 (1981). doi: 10.1007/BF01942327.

[154] L. Haslehurst and R. Penrose, The most general (2,2) self-dual vacuum, *Twist. Newsl.* **34**, 1 (1992).

[155] R. Penrose and G. A. J. Sparling, The twistor quadrille, *Twist. Newsl.* **1**, 10 (1976).

[156] J. Dalhuisen and D. Bouwmeester, Twistors and electromagnetic knots, *J. Phys. A* **45**, 135201 (2012). doi: 10.1088/1751-8113/45/ 13/135201.

[157] J. Swearngin, A. Thompson, A. Wickes, J. W. Dalhuisen and D. Bouwmeester, Gravitational hopfions (2013). arXiv:1302.1431 [gr-qc].

[158] A. J. de Klerk, R. I. van der Veen, J. W. Dalhuisen and D. Bouwmeester, Knotted optical vortices in exact solutions to Maxwell's equations, *Phys. Rev. A* **95**(5), 053820 (2017). doi: 10.1103/PhysRevA.95.053820, arXiv:1610.05285 [math-ph].

[159] A. Thompson, A. Wickes, J. Swearngin and D. Bouwmeester, Classification of electromagnetic and gravitational hopfions by algebraic type, *J. Phys. A* **48**(20), 205202 (2015). doi: 10.1088/1751-8113/48/20/205202, arXiv:1411.2073 [gr-qc].

[160] A. Thompson, J. Swearngin and D. Bouwmeester, Linked and knotted gravitational radiation, *J. Phys. A* **47**, 355205 (2014). doi: 10.1088/1751-8113/47/35/355205, arXiv:1402.3806 [gr-qc].

[161] T. Adamo and U. Kol, Classical double copy at null infinity, *Class. Quant. Grav.* **39**(10), 105007 (2022). doi: 10.1088/1361-6382/ac635e, arXiv:2109.07832 [hep-th].

[162] L. J. Mason, Dolbeault representatives from characteristic initial data at null infinity, *Twist. Newsl.* **22**, 28 (1986).

[163] N. M. J. Woodhouse, Real methods in twistor theory, *Class. Quant. Grav.* **2**, 257–291 (1985). doi: 10.1088/0264-9381/2/3/006.

[164] A. Luna, N. Moynihan and C. D. White, Why is the Weyl double copy local in position space? (2022). arXiv:2208.08548 [hep-th].

[165] D. A. Kosower, B. Maybee and D. O'Connell, Amplitudes, observables, and classical scattering, *JHEP* **02**, 137 (2019). doi: 10.1007/JHEP02(2019)137, arXiv:1811.10950 [hep-th].

[166] A. Guevara, Reconstructing classical spacetimes from the S-matrix in twistor space (2021). arXiv:2112.05111 [hep-th].

[167] D. A. Easson, T. Manton and A. Svesko, Sources in the Weyl double copy, *Phys. Rev. Lett.* **127**(27), 271101 (2021). doi: 10.1103/PhysRevLett.127.271101, arXiv:2110.02293 [gr-qc].

[168] D. A. Easson, T. Manton and A. Svesko, Einstein-Maxwell theory and the Weyl double copy (2022). arXiv:2210.16339 [gr-qc].

[169] K. Armstrong-Williams and C. D. White, A spinorial double copy for $\mathcal{N} = 0$ supergravity (2023). arXiv:2303.04631 [hep-th].

[170] A. Anastasiou, L. Borsten, M. J. Duff, L. J. Hughes and S. Nagy, Yang–Mills origin of gravitational symmetries, *Phys. Rev. Lett.* **113**(23), 231606 (2014). doi: 10.1103/PhysRevLett.113.231606, arXiv:1408.4434 [hep-th].

[171] A. Anastasiou, L. Borsten, M. J. Duff, M. J. Hughes, A. Marrani, S. Nagy and M. Zoccali, Twin supergravities from Yang–Mills theory squared, *Phys. Rev.* **D96**(2), 026013 (2017). doi: 10.1103/PhysRevD.96.026013, arXiv:1610.07192 [hep-th].

[172] G. L. Cardoso, S. Nagy and S. Nampuri, A double copy for $\mathcal{N} = 2$ supergravity: A linearised tale told on-shell, *JHEP* **10**, 127 (2016). doi: 10.1007/JHEP10(2016)127, arXiv:1609.05022 [hep-th].

[173] G. Cardoso, S. Nagy and S. Nampuri, Multi-centered $\mathcal{N} = 2$ BPS black holes: A double copy description, *JHEP* **04**, 037 (2017). doi: 10.1007/JHEP04(2017)037, arXiv:1611.04409 [hep-th].

[174] A. Anastasiou, L. Borsten, M. J. Duff, A. Marrani, S. Nagy and M. Zoccali, Are all supergravity theories Yang–Mills squared? (2017). arXiv:1707.03234 [hep-th].

[175] A. Anastasiou, L. Borsten, M. J. Duff, A. Marrani, S. Nagy and M. Zoccali, The mile high magic pyramid (2017). arXiv:1711.08476 [hep-th], https://inspirehep.net/record/1638344/files/arXiv:1711.08476.pdf.

[176] L. Borsten, M. J. Duff, L. J. Hughes and S. Nagy, Magic square from Yang–Mills squared, *Phys. Rev. Lett.* **112**(13), 131601 (2014). doi: 10.1103/PhysRevLett.112.131601, arXiv:1301.4176 [hep-th].

[177] A. Anastasiou, L. Borsten, M. J. Duff, L. J. Hughes and S. Nagy, A magic pyramid of supergravities, *JHEP* **04**, 178 (2014). doi: 10.1007/JHEP04(2014)178, arXiv:1312.6523 [hep-th].

[178] A. Anastasiou, L. Borsten, M. Hughes and S. Nagy, Global symmetries of Yang–Mills squared in various dimensions (2015). arXiv:1502.05359 [hep-th].

[179] I. A. Batalin and G. A. Vilkovisky, Relativistic S matrix of dynamical systems with boson and fermion constraints, *Phys. Lett.* **69B**, 309–312 (1977). doi: 10.1016/0370-2693(77)90553-6.

[180] E. S. Fradkin and G. A. Vilkovisky, Quantization of relativistic systems with constraints, *Phys. Lett.* **55B**, 224–226 (1975). doi: 10.1016/0370-2693(75)90448-7.

[181] I. A. Batalin and G. A. Vilkovisky, Gauge algebra and quantization, *Phys. Lett. B* **102**, 27–31 (1981). doi: 10.1016/0370-2693(81)90205-7.

[182] I. A. Batalin and G. A. Vilkovisky, Quantization of gauge theories with linearly dependent generators, *Phys. Rev. D* **28**, 2567–2582 (1983). doi: 10.1103/PhysRevD.28.2567. [Erratum: *Phys. Rev. D* **30**, 508 (1984)].

[183] I. A. Batalin and G. A. Vilkovisky, Closure of the gauge algebra, generalized lie equations and Feynman rules, *Nucl. Phys. B* **234**, 106–124 (1984). doi: 10.1016/0550-3213(84)90227-X.

[184] I. A. Batalin and G. A. Vilkovisky, Existence theorem for gauge algebra, *J. Math. Phys.* **26**, 172–184 (1985). doi: 10.1063/1.526780.

[185] C. Becchi, A. Rouet and R. Stora, The abelian Higgs-Kibble model. Unitarity of the S operator, *Phys. Lett.* **52B**, 344–346 (1974). doi: 10.1016/0370-2693(74)90058-6.

[186] C. Becchi, A. Rouet and R. Stora, Renormalization of the abelian Higgs-Kibble model, *Commun. Math. Phys.* **42**, 127–162 (1975). doi: 10.1007/BF01614158.

[187] C. Becchi, A. Rouet and R. Stora, Renormalization of gauge theories, *Ann. Phys.* **98**, 287–321 (1976). doi: 10.1016/0003-4916(76)90156-1.

[188] I. V. Tyutin, Gauge invariance in field theory and statistical physics in operator formalism (1975). arXiv:0812.0580 [hep-th].

[189] A. Anastasiou, L. Borsten, M. J. Duff, S. Nagy and M. Zoccali, Gravity as gauge theory squared: A ghost story, *Phys. Rev. Lett.* **121**(21), 211601 (2018). doi: 10.1103/PhysRevLett.121.211601, arXiv:1807.02486 [hep-th].

[190] L. Borsten, I. Jubb, V. Makwana and S. Nagy, Gauge × gauge on spheres, *JHEP* **06**, 096 (2020). doi: 10.1007/JHEP06(2020)096, arXiv:1911.12324 [hep-th].

[191] A. Luna, S. Nagy and C. White, The convolutional double copy: A case study with a point, *JHEP* **09**, 062 (2020). doi: 10.1007/JHEP09(2020)062, arXiv:2004.11254 [hep-th].

[192] M. Godazgar, C. N. Pope, A. Saha and H. Zhang, BRST symmetry and the convolutional double copy, *JHEP* **11**, 038 (2022). doi: 10.1007/JHEP11(2022)038, arXiv:2208.06903 [hep-th].

[193] L. Borsten and S. Nagy, The pure BRST Einstein-Hilbert Lagrangian from the double-copy to cubic order, *JHEP* **07**, 093 (2020). doi: 10.1007/JHEP07(2020)093, arXiv:2004.14945 [hep-th].

[194] C. D. White, Exact solutions for the biadjoint scalar field, *Phys. Lett.* **B763**, 365–369 (2016). doi: 10.1016/j.physletb.2016.10.052, arXiv:1606.04724 [hep-th].

[195] P.-J. De Smet and C. D. White, Extended solutions for the biadjoint scalar field, *Phys. Lett.* **B775**, 163–167 (2017). doi: 10.1016/j.physletb.2017.11.007, arXiv:1708.01103 [hep-th].

[196] N. Bahjat-Abbas, R. Stark-Muchão and C. D. White, Biadjoint wires, *Phys. Lett.* **B788**, 274–279 (2019). doi: 10.1016/j.physletb.2018.11.026, arXiv:1810.08118 [hep-th].

[197] A. Banerjee, E. Colgáin, J. A. Rosabal and H. Yavartanoo, Ehlers as EM duality in the double copy (2019). arXiv:1912.02597 [hep-th].

[198] Y.-T. Huang, U. Kol and D. O'Connell, The double copy of electric-magnetic duality (2019). arXiv:1911.06318 [hep-th].

[199] D. S. Berman, E. Chacón, A. Luna and C. D. White, The self-dual classical double copy, and the Eguchi-Hanson instanton (2018). arXiv:1809.04063 [hep-th].

[200] L. Alfonsi, C. D. White and S. Wikeley, Topology and Wilson lines: Global aspects of the double copy, *JHEP* **07**, 091 (2020). doi: 10.1007/JHEP07(2020)091, arXiv:2004.07181 [hep-th].

[201] R. Alawadhi, D. S. Berman, C. D. White and S. Wikeley, The single copy of the gravitational holonomy (2021). arXiv:2107.01114 [hep-th].

[202] C. Cheung, J. Mangan, J. Parra-Martinez and N. Shah, Nonperturbative double copy in Flatland (2022). arXiv:2204.07130 [hep-th].

[203] K. Armstrong-Williams, C. D. White and S. Wikeley, Nonperturbative aspects of the self-dual double copy, *JHEP* **08**, 160 (2022). doi: 10.1007/JHEP08(2022)160, arXiv:2205.02136 [hep-th].

[204] J. Plebański, Some solutions of complex Einstein equations, *J. Math. Phys.* **16**, 2395–2402 (1975). doi: 10.1063/1.522505.

[205] A. Parkes, A cubic action for selfdual Yang–Mills, *Phys. Lett.* **B286**, 265–270 (1992). doi: 10.1016/0370-2693(92)91773-3, arXiv:hep-th/9203074 [hep-th].

[206] B. F. Schutz, *Geometrical Methods of Mathematical Physics*. Cambridge University Press, Cambridge, UK (1980). doi: 10.1017/CBO9781139171540.

[207] E. Chacón, H. García-Compeán, A. Luna, R. Monteiro and C. D. White, New heavenly double copies, *JHEP* **03**, 247 (2021). doi: 10.1007/JHEP03(2021)247, arXiv:2008.09603 [hep-th].

[208] J. Hoppe, Diffeomorphism groups, quantization and SU(infinity), *Int. J. Mod. Phys. A* **4**, 5235 (1989). doi: 10.1142/S0217751X8900 2235.

[209] M. Ben-Shahar and H. Johansson, Off-shell color-kinematics duality for Chern-Simons, *JHEP* **08**, 035 (2022). doi: 10.1007/ JHEP08(2022)035, arXiv:2112.11452 [hep-th].

[210] L. Borsten, B. Jurco, H. Kim, T. Macrelli, C. Saemann and M. Wolf, Becchi-Rouet-Stora-Tyutin-Lagrangian double copy of Yang–Mills theory, *Phys. Rev. Lett.* **126**(19), 191601 (2021). doi: 10.1103/ PhysRevLett.126.191601, arXiv:2007.13803 [hep-th].

[211] L. Borsten, H. Kim, B. Jurčo, T. Macrelli, C. Saemann and M. Wolf, Tree-level color–kinematics duality implies loop-level color–kinematics duality up to counterterms, *Nucl. Phys. B* **989**, 116144 (2023). doi: 10.1016/j.nuclphysb.2023.116144, arXiv:2108. 03030 [hep-th].

[212] L. Borsten, H. Kim, B. Jurco, T. Macrelli, C. Saemann and M. Wolf, Colour-kinematics duality, double copy, and homotopy algebras, *PoS* **ICHEP2022**, 426 (2022). doi: 10.22323/1.414.0426, arXiv:2211.16405 [hep-th].

[213] L. Borsten, B. Jurco, H. Kim, T. Macrelli, C. Saemann and M. Wolf, Kinematic lie algebras from twistor spaces (2022). arXiv:2211.13261 [hep-th].

[214] L. Borsten, B. Jurco, H. Kim, T. Macrelli, C. Saemann and M. Wolf, Double copy from tensor products of metric BV-algebras (2023). arXiv:2307.02563 [hep-th].

[215] F. Diaz-Jaramillo, O. Hohm and J. Plefka, Double field theory as the double copy of Yang–Mills theory, *Phys. Rev. D* **105**(4), 045012 (2022). doi: 10.1103/PhysRevD.105.045012, arXiv:2109.01153 [hep-th].

[216] R. Bonezzi, F. Diaz-Jaramillo and O. Hohm, The gauge structure of double field theory follows from Yang–Mills theory, *Phys. Rev. D* **106**(2), 026004 (2022). doi: 10.1103/PhysRevD.106.026004, arXiv:2203.07397 [hep-th].

[217] R. Bonezzi, C. Chiaffrino, F. Diaz-Jaramillo and O. Hohm, Gauge invariant double copy of Yang–Mills theory: The quartic theory (2022). arXiv:2212.04513 [hep-th].

[218] R. Bonezzi, C. Chiaffrino, F. Diaz-Jaramillo and O. Hohm, Gravity = Yang–Mills (2023). arXiv:2306.14788 [hep-th].

[219] S. Mizera, Kinematic Jacobi identity is a residue theorem: Geometry of color-kinematics duality for gauge and gravity amplitudes, *Phys. Rev. Lett.* **124**(14), 141601 (2020). doi: 10.1103/PhysRevLett.124.141601, arXiv:1912.03397 [hep-th].

[220] C.-H. Fu and K. Krasnov, Colour-kinematics duality and the Drinfeld double of the Lie algebra of diffeomorphisms, *JHEP* **01**, 075 (2017). doi: 10.1007/JHEP01(2017)075, arXiv:1603.02033 [hep-th].

[221] G. Chen, H. Johansson, F. Teng and T. Wang, On the kinematic algebra for BCJ numerators beyond the MHV sector, *JHEP* **11**, 055 (2019). doi: 10.1007/JHEP11(2019)055, arXiv:1906.10683 [hep-th].

[222] G. Chen, G. Lin and C. Wen, Kinematic Hopf algebra for amplitudes and form factors (2022). arXiv:2208.05519 [hep-th].

[223] A. Brandhuber, G. Chen, H. Johansson, G. Travaglini and C. Wen, Kinematic Hopf algebra for Bern-Carrasco-Johansson numerators in heavy-mass effective field theory and Yang–Mills theory, *Phys. Rev. Lett.* **128**(12), 121601 (2022). doi: 10.1103/PhysRevLett.128.121601, arXiv:2111.15649 [hep-th].

[224] A. Brandhuber, G. R. Brown, G. Chen, J. Gowdy, G. Travaglini and C. Wen, Amplitudes, Hopf algebras and the colour-kinematics duality, *JHEP* **12**, 101 (2022). doi: 10.1007/JHEP12(2022)101, arXiv:2208.05886 [hep-th].

[225] C. R. Mafra and O. Schlotterer, Multiparticle SYM equations of motion and pure spinor BRST blocks, *JHEP* **1407**, 153 (2014). doi: 10.1007/JHEP07(2014)153, arXiv:1404.4986 [hep-th].

[226] C. R. Mafra and O. Schlotterer, Berends-Giele recursions and the BCJ duality in superspace and components, *JHEP* **03**, 097 (2016). doi: 10.1007/JHEP03(2016)097, arXiv:1510.08846 [hep-th].

[227] E. J. Weinberg, *Classical Solutions in Quantum Field Theory*. Cambridge Monographs on Mathematical Physics. Cambridge University

Press, Cambridge, UK (2012). doi: 10.1017/CBO9781139017787,
http://www.cambridge.org/us/knowledge/isbn/item6813336/.

[228] N. S. Manton and P. Sutcliffe, *Topological Solitons*. Cambridge
       Monographs on Mathematical Physics. Cambridge University Press,
       Cambridge, UK (2004). doi: 10.1017/CBO9780511617034, http://
       www.cambridge.org/uk/catalogue/catalogue.asp?isbn=0521838363.

[229] C.-N. Yang, *Selected Papers (1945–1980) Of Chen Ning Yang
       (With Commentary)*. 36 (World Scientific Series In 20th Cen-
       tury Physics). World Scientific, Singapore (2005). https://www.
       amazon.co.uk/Selected-1945-1980-Commentary-Scientific-Century/
       dp/9812563679.

[230] G. H. Derrick, Comments on nonlinear wave equations as models
       for elementary particles, *J. Math. Phys.* **5**, 1252–1254 (1964). doi:
       10.1063/1.1704233.

[231] T. Eguchi and A. J. Hanson, Asymptotically flat selfdual solutions
       to euclidean gravity, *Phys. Lett.* **74B**, 249–251 (1978). doi: 10.1016/
       0370-2693(78)90566-X.

[232] T. Eguchi and A. J. Hanson, Selfdual solutions to euclidean gravity,
       Ann. Phys. **120**, 82 (1979). doi: 10.1016/0003-4916(79)90282-3.

[233] T. Eguchi and A. J. Hanson, Gravitational instantons, *Gen. Rel.
       Grav.* **11**, 315–320 (1979). doi: 10.1007/BF00759271.

[234] J. A. Cronin, Phenomenological model of strong and weak interac-
       tions in chiral U(3) x U(3), *Phys. Rev.* **161**, 1483–1494 (1967). doi:
       10.1103/PhysRev.161.1483.

[235] S. Weinberg, Dynamical approach to current algebra, *Phys. Rev.
       Lett.* **18**, 188–191 (1967). doi: 10.1103/PhysRevLett.18.188.

[236] S. Weinberg, Nonlinear realizations of chiral symmetry, *Phys. Rev.*
       166, 1568–1577 (1968). doi: 10.1103/PhysRev.166.1568.

[237] M. Born and L. Infeld, Foundations of the new field theory, *Proc.
       R. Soc. Lond. A* **144**(852), 425–451 (1934). doi: 10.1098/rspa.1934.
       0059.

[238] M. Born, On the quantum theory of the electromagnetic field, *Proc.
       R. Soc. Lond. A* **143**(849), 410–437 (1934). doi: 10.1098/rspa.1934.
       0010.

[239] M. Born, Nonlinear theory of the electromagnetic field, *Ann. Inst.
       Henri Poincare* **7**(4), 155–265 (1937).

[240] J. Rafelski, G. Soff and W. Greiner, Lower bound to limiting fields
       in nonlinear electrodynamics, *Phys. Rev. A* **7**, 903 (1973). doi: 10.
       1103/PhysRevA.7.903.

[241] E. S. Fradkin and A. A. Tseytlin, Nonlinear electrodynamics from
       quantized strings, *Phys. Lett. B* **163**, 123–130 (1985). doi: 10.1016/
       0370-2693(85)90205-9.

[242] R. G. Leigh, Dirac-Born-Infeld action from Dirichlet sigma model, *Mod. Phys. Lett. A* **4**, 2767 (1989). doi: 10.1142/S0217732389003099.

[243] K. Becker, M. Becker and J. H. Schwarz, *String Theory and M-Theory: A Modern Introduction*. Cambridge University Press, Cambridge, UK (2006). doi: 10.1017/CBO9780511816086.

[244] J. Bagger and A. Galperin, A new Goldstone multiplet for partially broken supersymmetry, *Phys. Rev. D* **55**, 1091–1098 (1997). doi: 10.1103/PhysRevD.55.1091, arXiv:hep-th/9608177.

[245] E. Bergshoeff, F. Coomans, R. Kallosh, C. S. Shahbazi and A. Van Proeyen, Dirac-Born-Infeld-Volkov-Akulov and deformation of supersymmetry, *JHEP* **08**, 100 (2013). doi: 10.1007/JHEP08(2013)100, arXiv:1303.5662 [hep-th].

[246] A. Nicolis, R. Rattazzi and E. Trincherini, The Galileon as a local modification of gravity, *Phys. Rev. D* **79**, 064036 (2009). doi: 10.1103/PhysRevD.79.064036, arXiv:0811.2197 [hep-th].

[247] G. R. Dvali, G. Gabadadze and M. Porrati, 4-D gravity on a brane in 5-D Minkowski space, *Phys. Lett. B* **485**, 208–214 (2000). doi: 10.1016/S0370-2693(00)00669-9, arXiv:hep-th/0005016.

[248] C. de Rham, G. Gabadadze and A. J. Tolley, Resummation of massive gravity, *Phys. Rev. Lett.* **106**, 231101 (2011). doi: 10.1103/PhysRevLett.106.231101, arXiv:1011.1232 [hep-th].

[249] C. de Rham and A. J. Tolley, DBI and the Galileon reunited, *JCAP* **05**, 015 (2010). doi: 10.1088/1475-7516/2010/05/015, arXiv:1003.5917 [hep-th].

[250] C. Cheung, K. Kampf, J. Novotny and J. Trnka, Effective field theories from soft limits of scattering amplitudes, *Phys. Rev. Lett.* **114**(22), 221602 (2015). doi: 10.1103/PhysRevLett.114.221602, arXiv:1412.4095 [hep-th].

[251] F. Cachazo, S. He and E. Y. Yuan, Scattering equations and matrices: From Einstein to Yang–Mills, DBI and NLSM, *JHEP* **07**, 149 (2015). doi: 10.1007/JHEP07(2015)149, arXiv:1412.3479 [hep-th].

[252] K. Hinterbichler and A. Joyce, Hidden symmetry of the Galileon, *Phys. Rev. D* **92**(2), 023503 (2015). doi: 10.1103/PhysRevD.92.023503, arXiv:1501.07600 [hep-th].

[253] J. Novotny, Geometry of special Galileons, *Phys. Rev. D* **95**(6), 065019 (2017). doi: 10.1103/PhysRevD.95.065019, arXiv:1612.01738 [hep-th].

[254] F. Cachazo, S. He and E. Y. Yuan, Scattering of massless particles: Scalars, gluons and gravitons, *JHEP* **07**, 033 (2014). doi: 10.1007/JHEP07(2014)033, arXiv:1309.0885 [hep-th].

[255] F. Cachazo, S. He and E. Y. Yuan, Scattering of massless particles in arbitrary dimensions, *Phys. Rev. Lett.* **113**(17), 171601 (2014). doi: 10.1103/PhysRevLett.113.171601, arXiv:1307.2199 [hep-th].

[256] D. B. Fairlie and D. E. Roberts, Dual models without tachyons – A new approach, Print-72-2440 (1972).

[257] T. Adamo, J. J. M. Carrasco, M. Carrillo-González, M. Chiodaroli, H. Elvang, H. Johansson, D. O'Connell, R. Roiban and O. Schlotterer, Snowmass white paper: The double copy and its applications. In *2022 Snowmass Summer Study* (2022). arXiv:2204.06547 [hep-th].

[258] L. Mason and D. Skinner, Ambitwistor strings and the scattering equations, *JHEP* **1407**, 048 (2014). doi: 10.1007/JHEP07(2014)048, arXiv:1311.2564 [hep-th].

[259] C. Cheung and C.-H. Shen, Symmetry for flavor-kinematics duality from an action, *Phys. Rev. Lett.* **118**(12), 121601 (2017). doi: 10.1103/PhysRevLett.118.121601, arXiv:1612.00868 [hep-th].

[260] C. Cheung, A. Helset and J. Parra-Martinez, Geometry-kinematics duality, *Phys. Rev. D* **106**(4), 045016 (2022). doi: 10.1103/PhysRevD.106.045016, arXiv:2202.06972 [hep-th].

[261] C. Cheung and J. Mangan, Covariant color-kinematics duality, *JHEP* **11**, 069 (2021). doi: 10.1007/JHEP11(2021)069, arXiv:2108.02276 [hep-th].

[262] S. He, F. Teng and Y. Zhang, String amplitudes from field-theory amplitudes and vice versa, *Phys. Rev. Lett.* **122**(21), 211603 (2019). doi: 10.1103/PhysRevLett.122.211603, arXiv:1812.03369 [hep-th].

[263] T. Azevedo, M. Chiodaroli, H. Johansson and O. Schlotterer, Heterotic and bosonic string amplitudes via field theory, *JHEP* **10**, 012 (2018). doi: 10.1007/JHEP10(2018)012, arXiv:1803.05452 [hep-th].

[264] M. Duff, Quantum tree graphs and the Schwarzschild solution, *Phys. Rev.* **D7**, 2317–2326 (1973). doi: 10.1103/PhysRevD.7.2317.

[265] W. D. Goldberger and A. K. Ridgway, Radiation and the classical double copy for color charges, *Phys. Rev.* **D95**(12), 125010 (2017). doi: 10.1103/PhysRevD.95.125010, arXiv:1611.03493 [hep-th].

[266] A. Luna, R. Monteiro, I. Nicholson, A. Ochirov, D. O'Connell, N. Westerberg and C. D. White, Perturbative spacetimes from Yang–Mills theory, *JHEP* **04**, 069 (2017). doi: 10.1007/JHEP04(2017)069, arXiv:1611.07508 [hep-th].

[267] A. I. Janis, E. T. Newman and J. Winicour, Reality of the Schwarzschild singularity, *Phys. Rev. Lett.* **20**, 878–880 (1968). doi: 10.1103/PhysRevLett.20.878.

[268] K. Kim, K. Lee, R. Monteiro, I. Nicholson and D. Peinador Veiga, The classical double copy of a point charge, *JHEP* **02**, 046 (2020). doi: 10.1007/JHEP02(2020)046, arXiv:1912.02177 [hep-th].

[269] C.-H. Shen, Gravitational radiation from color-kinematics duality, *JHEP* **11**, 162 (2018). doi: 10.1007/JHEP11(2018)162, arXiv:1806.07388 [hep-th].

[270] W. D. Goldberger, S. G. Prabhu and J. O. Thompson, Classical gluon and graviton radiation from the bi-adjoint scalar double copy, *Phys. Rev.* **D96**(6), 065009 (2017). doi: 10.1103/PhysRevD. 96.065009, arXiv:1705.09263 [hep-th].

[271] W. D. Goldberger and A. K. Ridgway, Bound states and the classical double copy, *Phys. Rev.* **D97**(8), 085019 (2018). doi: 10.1103/PhysRevD.97.085019, arXiv:1711.09493 [hep-th].

[272] W. D. Goldberger, J. Li and S. G. Prabhu, Spinning particles, axion radiation, and the classical double copy, *Phys. Rev.* **D97**(10), 105018 (2018). doi: 10.1103/PhysRevD.97.105018, arXiv:1712.09250 [hep-th].

[273] J. Li and S. G. Prabhu, Gravitational radiation from the classical spinning double copy, *Phys. Rev. D* **97**(10), 105019 (2018). doi: 10.1103/PhysRevD.97.105019, arXiv:1803.02405 [hep-th].

[274] D. Chester, Radiative double copy for Einstein-Yang–Mills theory, *Phys. Rev. D* **97**(8), 084025 (2018). doi: 10.1103/PhysRevD.97. 084025, arXiv:1712.08684 [hep-th].

[275] A. Luna, I. Nicholson, D. O'Connell and C. D. White, Inelastic black hole scattering from charged scalar amplitudes, *JHEP* **03**, 044 (2018). doi: 10.1007/JHEP03(2018)044, arXiv:1711.03901 [hep-th].

[276] H. Johansson and A. Ochirov, Pure gravities via color-kinematics duality for fundamental matter, *JHEP* **11**, 046 (2015). doi: 10.1007/ JHEP11(2015)046, arXiv:1407.4772 [hep-th].

[277] R. A. Hulse and J. H. Taylor, Discovery of a pulsar in a binary system, *Astrophys. J. Lett.* **195**, L51–L53 (1975). doi: 10.1086/181708.

[278] J. H. Taylor and J. M. Weisberg, A new test of general relativity: Gravitational radiation and the binary pulsar PS R 1913+16, *Astrophys. J.* **253**, 908–920 (1982). doi: 10.1086/159690.

[279] B. P. Abbott *et al.*, Observation of gravitational waves from a binary black hole merger, *Phys. Rev. Lett.* **116**(6), 061102 (2016). doi: 10.1103/PhysRevLett.116.061102, arXiv:1602.03837 [gr-qc].

[280] B. P. Abbott *et al.*, GW151226: Observation of gravitational waves from a 22-solar-mass binary black hole coalescence, *Phys. Rev. Lett.* **116**(24), 241103 (2016). doi: 10.1103/PhysRevLett.116.241103, arXiv:1606.04855 [gr-qc].

[281] B. P. Abbott *et al.*, GW170814: A three-detector observation of gravitational waves from a binary black hole coalescence, *Phys. Rev.*

Lett. **119**(14), 141101 (2017). doi: 10.1103/PhysRevLett.119.141101, arXiv:1709.09660 [gr-qc].

[282] B. P. Abbott *et al.*, GW170817: Observation of gravitational waves from a binary neutron star inspiral, *Phys. Rev. Lett.* **119**(16), 161101 (2017). doi: 10.1103/PhysRevLett.119.161101, arXiv:1710.05832 [gr-qc].

[283] M. Maggiore, *Gravitational Waves. Vol. 1: Theory and Experiments.* Oxford University Press, Oxford, UK (2007). doi: 10.1093/acprof: oso/9780198570745.001.0001.

[284] M. Maggiore, *Gravitational Waves. Vol. 2: Astrophysics and Cosmology.* Oxford University Press, Oxford, UK (2018).

[285] A. Buonanno, M. Khalil, D. O'Connell, R. Roiban, M. P. Solon and M. Zeng, Snowmass white paper: Gravitational waves and scattering amplitudes. In *Snowmass 2021* (2022). arXiv:2204.05194 [hep-th].

[286] J. Plefka, J. Steinhoff and W. Wormsbecher, Effective action of dilaton gravity as the classical double copy of Yang–Mills theory, *Phys. Rev. D* **99**(2), 024021 (2019). doi: 10.1103/PhysRevD.99.024021, arXiv:1807.09859 [hep-th].

[287] J. Plefka, C. Shi, J. Steinhoff and T. Wang, Breakdown of the classical double copy for the effective action of dilaton-gravity at NNLO (2019). arXiv:1906.05875 [hep-th].

[288] G. Mogull, J. Plefka and J. Steinhoff, Classical black hole scattering from a worldline quantum field theory, *JHEP* **02**, 048 (2021). doi: 10.1007/JHEP02(2021)048, arXiv:2010.02865 [hep-th].

[289] G. U. Jakobsen, G. Mogull, J. Plefka and J. Steinhoff, SUSY in the sky with gravitons, *JHEP* **01**, 027 (2022). doi: 10.1007/ JHEP01(2022)027, arXiv:2109.04465 [hep-th].

[290] C. Shi and J. Plefka, Classical double copy of worldline quantum field theory, *Phys. Rev. D* **105**(2), 026007 (2022). doi: 10.1103/ PhysRevD.105.026007, arXiv:2109.10345 [hep-th].

[291] A. P. Balachandran, P. Salomonson, B.-S. Skagerstam and J.-O. Winnberg, Classical description of particle interacting with non-abelian gauge field, *Phys. Rev. D* **15**, 2308–2317 (1977). doi: 10.1103/PhysRevD.15.2308.

[292] F. Comberiati and C. Shi, Classical double copy of spinning worldline quantum field theory, *J. High Energ. Phys.* **2023**, 8 (2023). https://doi.org/10.1007/JHEP04(2023)008.

[293] W. D. Goldberger and I. Z. Rothstein, An effective field theory of gravity for extended objects, *Phys. Rev. D* **73**, 104029 (2006). doi: 10.1103/PhysRevD.73.104029, arXiv:hep-th/0409156.

[294] M. Levi, Effective field theories of post-newtonian gravity: A comprehensive review, *Rep. Prog. Phys.* **83**(7), 075901 (2020). doi: 10.1088/1361-6633/ab12bc, arXiv:1807.01699 [hep-th].

[295] R. A. Porto, The effective field theorist's approach to gravitational dynamics, *Phys. Rep.* **633**, 1–104 (2016). doi: 10.1016/j.physrep. 2016.04.003, arXiv:1601.04914 [hep-th].

[296] A. Cristofoli, R. Gonzo, D. A. Kosower and D. O'Connell, Waveforms from amplitudes, *Phys. Rev. D* **106**(5), 056007 (2022). doi: 10.1103/PhysRevD.106.056007, arXiv:2107.10193 [hep-th].

[297] M. Froissart, Asymptotic behavior and subtractions in the Mandelstam representation, *Phys. Rev.* **123**, 1053–1057 (1961). doi: 10.1103/PhysRev.123.1053.

[298] M. Levy and J. Sucher, Eikonal approximation in quantum field theory, *Phys. Rev.* **186**, 1656–1670 (1969). doi: 10.1103/PhysRev. 186.1656.

[299] H. D. I. Abarbanel and C. Itzykson, Relativistic eikonal expansion, *Phys. Rev. Lett.* **23**, 53 (1969). doi: 10.1103/PhysRevLett. 23.53.

[300] H. Cheng and T. T. Wu, High-energy elastic scattering in quantum electrodynamics, *Phys. Rev. Lett.* **22**, 666 (1969). doi: 10.1103/ PhysRevLett.22.666.

[301] D. Amati, M. Ciafaloni and G. Veneziano, Classical and quantum gravity effects from planckian energy superstring collisions, *Int. J. Mod. Phys. A* **3**, 1615–1661 (1988). doi: 10.1142/ S0217751X88000710.

[302] D. Amati, M. Ciafaloni and G. Veneziano, Higher order gravitational deflection and soft bremsstrahlung in planckian energy superstring collisions, *Nucl. Phys. B* **347**, 550–580 (1990). doi: 10.1016/0550-3213(90)90375-N.

[303] D. N. Kabat and M. Ortiz, Eikonal quantum gravity and planckian scattering, *Nucl. Phys.* **B388**, 570–592 (1992). doi: 10.1016/ 0550-3213(92)90627-N, arXiv:hep-th/9203082 [hep-th].

[304] S. B. Giddings, M. Schmidt-Sommerfeld and J. R. Andersen, High energy scattering in gravity and supergravity, *Phys. Rev.* **D82**, 104022 (2010). doi: 10.1103/PhysRevD.82.104022, arXiv:1005.5408 [hep-th].

[305] G. D'Appollonio, P. Di Vecchia, R. Russo and G. Veneziano, High-energy string-brane scattering: Leading eikonal and beyond, *JHEP* **11**, 100 (2010). doi: 10.1007/JHEP11(2010)100, arXiv:1008.4773 [hep-th].

[306] P. Di Vecchia, A. Luna, S. G. Naculich, R. Russo, G. Veneziano and C. D. White, A tale of two exponentiations in $\mathcal{N} = 8$ supergravity, *Phys. Lett. B* **798**, 134927 (2019). doi: 10.1016/j.physletb.2019. 134927, arXiv:1908.05603 [hep-th].

[307] P. Di Vecchia, S. G. Naculich, R. Russo, G. Veneziano and C. D. White, A tale of two exponentiations in $\mathcal{N} = 8$ supergravity at

subleading level, *JHEP* **03**, 173 (2020). doi: 10.1007/JHEP03(2020)
173, arXiv:1911.11716 [hep-th].

[308]   P. Di Vecchia, C. Heissenberg, R. Russo and G. Veneziano,
Universality of ultra-relativistic gravitational scattering, *Phys.
Lett. B* **811**, 135924 (2020). doi: 10.1016/j.physletb.2020.135924,
arXiv:2008.12743 [hep-th].

[309]   P. Di Vecchia, C. Heissenberg, R. Russo and G. Veneziano,
The eikonal approach to gravitational scattering and radiation
at $\mathcal{O}(\mathrm{G}^3)$, *JHEP* **07**, 169 (2021). doi: 10.1007/JHEP07(2021)169,
arXiv:2104.03256 [hep-th].

[310]   P. Di Vecchia, C. Heissenberg, R. Russo and G. Veneziano,
The eikonal operator at arbitrary velocities I: The soft-radiation
limit, *JHEP* **07**, 039 (2022). doi: 10.1007/JHEP07(2022)039,
arXiv:2204.02378 [hep-th].

[311]   P. Di Vecchia, C. Heissenberg, R. Russo and G. Veneziano, Classical gravitational observables from the eikonal operator (2022).
arXiv:2210.12118 [hep-th].

[312]   Z. Bern, J. P. Gatica, E. Herrmann, A. Luna and M. Zeng, Scalar
QED as a toy model for higher-order effects in classical gravitational
scattering, *JHEP* **08**, 131 (2022). doi: 10.1007/JHEP08(2022)131,
arXiv:2112.12243 [hep-th].

[313]   R. Akhoury, R. Saotome and G. Sterman, High energy scattering
in perturbative quantum gravity at next to leading power, *Phys.
Rev. D* **103**(6), 064036 (2021). doi: 10.1103/PhysRevD.103.064036,
arXiv:1308.5204 [hep-th].

[314]   A. Brandhuber, G. Chen, G. Travaglini and C. Wen, A new gauge-
invariant double copy for heavy-mass effective theory, *JHEP* **07**, 047
(2021). doi: 10.1007/JHEP07(2021)047, arXiv:2104.11206 [hep-th].

[315]   A. Brandhuber, G. Chen, G. Travaglini and C. Wen, Classical
gravitational scattering from a gauge-invariant double copy, *JHEP*
10, 118 (2021). doi: 10.1007/JHEP10(2021)118, arXiv:2108.04216
[hep-th].

[316]   A. Brandhuber, G. R. Brown, G. Chen, S. De Angelis, J. Gowdy
and G. Travaglini, One-loop gravitational bremsstrahlung and waveforms from a heavy-mass effective field theory (2023). arXiv:2303.
06111 [hep-th].

[317]   P. H. Damgaard, L. Plante and P. Vanhove, On an exponential representation of the gravitational S-matrix, *JHEP* **11**, 213 (2021). doi:
10.1007/JHEP11(2021)213, arXiv:2107.12891 [hep-th].

[318]   I. Fernandez-Corbaton, M. Cirio, A. Büse, L. Lamata, E. Solano
and G. Molina-Terriza, Quantum emulation of gravitational waves,
Sci. Rep. **5**, 11538 (2015). doi: 10.1038/srep11538, arXiv:1406.4263
[quant-ph].

Index

Printed in the USA
CPSIA information can be obtained
at www.ICGtesting.com
LVHW070045090524
779191LV00002B/150

9 781800 615458